THE COMPLETE GUIDE TO

HOME ROOFING INSTALLATION & MAINTENANCE

Text photographs by John Chiles
Prepress services by Studio 500 Associates
Cover photo compliments of
Owens-Corning: Crestwood Shadow — Shale

97 96 95 94   5 4 3 2

Library of Congress Cataloging-in-Publication Data

Chiles, John W., 1945-
    The complete guide to home roofing installation and maintenance : how to do it yourself and avoid the 60 ways your roofer can nail you / John W. Chiles, Jr. — 1st ed.
        p.   cm.
    Includes index.
    ISBN 1-55870-277-6
    1. Roofing — Amaeturs' manuals. 2. Roofs — Maintenance and repair — Amateurs' manuals.   I. title.
    TH2431.C48   1993
    695 — dc20

                                                                92-38663
                                                                CIP

This book is dedicated to my wife, Judy, who truly is the wind beneath my wings
and
My daughter, Lisa, who climbed through a window with us when the door slammed
and
My son, Daniel, whose eyes always see a hero.

Acknowledgments

Many thanks to Jerry Neri for the use of his boom truck and to Ed Meshyock, Jerry's tree foreman, who was very nervous about operating the camera and totally unconcerned about swinging through the air in a "bucket" fifty-five feet above the ground.

Contents

Introduction

YOU CAN DO IT! This book will share with you not only the "how," but the "why" of roofing with fiberglass asphalt shingles. In addition to the straightforward technical details of laying the shingles, this book gives helpful pointers that will save you labor and enhance the quality of your new roof. Much of the language you will need to know as well as some anecdotes, which help define what the roofing business is all about, are woven throughout the book.

The roof is the second most critical structural component of a building. (The foundation is the most critical.) Your roof is a crucial architectural element, and its color and texture are key features in the appearance of your home. The color of the shingles you select affects the durability of the shingles themselves, energy efficiency (future costs), and the comfort level of your home. The quality and finish of your roof will also have a dramatic effect on your home's future marketability.

Your roof is a major investment of money and time — whether you contract for the work or do it yourself. Laying a roof is so labor intensive that it can almost be classified as a "handcraft." This means that the details of the installation are extremely important. A properly installed roof is something you will not think about again for years. A bad roofing job, however, can cause major damage and will be aggravating, time consuming, and expensive to fix. In some cases, the homeowner — you — may be forced to redo the work completely. Don't worry. Whether you decide to do it yourself or contract the work, this book gives you the knowledge and insight you need to succeed. YOU CAN DO IT!

The average homeowner knows very little about the roof over his head. Lack of knowledge or incomplete understanding always means trouble for an eager do-it-yourselfer. Even if a homeowner is dealing with a reputable contractor, not knowing the basics about roofing may cause misunderstandings and tension. The worst possible case is a susceptible homeowner contracting with a disreputable contractor. Someday scientific studies will prove that vulnerable homeowners give off a scent that draws the jackals of the trades.

Quality contractors take pride in their work and want their customers to at least partially understand and appreciate what goes into it. Quality contractors don't want to deal with a homeowner who says, "A roof's a roof; they're all the same."

My customers were interested in my work. I was always glad to show them what I was doing and explain why I was doing it. Many told me I should write a book about roofing. For so many years customers took my advice and recommendations; now it seems logical that I accept theirs.

The first question I faced was, "What level of expertise will my reader have?" Too many times I have experienced the breakdown of the word "assume" into its three parts. The word "assume" makes an "ASS/ (of) U/ (and) ME." I decided it was best to accept the risk of over-explaining things. I wrote this book assuming you know nothing about roofing.

A word of caution about roofing is needed. Understanding roofing is not hard, but roofing is hard.

Most roofs can be done in sections, and I will show you how to control the work required on any given day. However, pitched (sloped) surfaces place an unaccustomed strain on your feet, ankles, knees, and lower back. Please be realistic about your physical capabilities and limitations. You don't want to tear off half of your roof then realize that you are too exhausted even to climb down the ladder, much less lay the new roof.

Every page will show you that you really can do it, *but* there is something I must say in the very beginning. I'm going to show you how to use roofing jacks and scaffold on your roof, but I don't want anyone getting hurt. If this is the first roof you've ever done and if your roof is steeper than a 6/12 (6 inches of vertical *rise* for every 12 inches of horizontal *run*), leave it to a professional. When a roof is steeper than 6/12, you can slip at the peak or ridge, claw and struggle as you slide down the shingles and pick up speed all the way to the ground. No fall on a roof is a good fall; there are only falls where you get hurt less than you might have. When the pitch is over 6/12, the risks far outweigh the potential for savings and sense of personal achievement.

I should also caution those of you who are planning to contract the work. The methods I use give perfectionist results. Most roofing companies don't do this kind of finished work. This doesn't mean the company is doing shoddy work on your roof. If I've done my job well, you will end up with a sound, fair contract with a quality contractor and you will know the difference between major "cheats" and minor "variations."

Let's address your basic question first. Do you contract the work or do you do it yourself? In the first chapter I am going to discuss contracting. Whichever direction you go, there will be a final exam during the first driving rainstorm.

1.
Contracting

There are many outstanding roofing companies, and your job, if you contract, is to find the right company. Your roof is a long-term investment spanning twenty to thirty-five years. This places you in the position of dealing with a contractor who knows you won't need his services again for twenty years or longer. Once you have paid your contractor for completing your roof, the only residual value he can gain from you is the referral work you send him. A top quality contractor lives by referrals and will make sure his customers are pleased.

The flip side of this is that once you have paid your contractor for completing your roof, the only residual value he can gain from you is the referral work you send him. A disreputable contractor doesn't care about references from you. There are enough easy marks out there that he doesn't give a rip about anything but slapping your roof on as quickly as possible and getting your money. Things could be worse. Your contractor could take your deposit and disappear forever. Things could be worse still. He could screw up the job so badly that you suffer extensive damage to your home — especially the interior.

Throughout this book I have inserted specific words of caution highlighted by the word "**Nail**". These "nails" include many of the shortcuts and the reasons bad contractors use them. The basic reasons all boil down to time and money.

Don't panic. I just want you to receive full value for your money and effort. There are several things you

can do to make sure you get the right contractor.

1. If a friend tells you he is pleased with the contractor who did his roof, ask questions. Did the contractor keep him advised regarding his schedule? Did the work proceed on schedule? Were shrubs and flower borders protected and kept clean during the job? Is the completed work neat? Are the gutters and downspouts clean? Did they receive a warranty from the contractor and the manufacturer's 20-, 25-, or 30-year warranty on the shingles themselves? If the answers are all yes, get that contractor's name and number and call him.

2. Talk to the manager of the local roofing materials supplier and ask for the names of contractors who are doing top quality work.

3. Check with state and county offices to makes sure your contractor is properly licensed. Also check with the Better Business Bureau to see if they have knowledge, good or bad, about the contractors you are considering.

4. Many established companies are listed with Dun & Bradstreet. Ask for the DUNS Number, but be prepared — they may not want to give this to you. Dun & Bradstreet gives each company a rating, but you probably can't get it unless you are a member of Dun & Bradstreet too. The rating gives the range of their annual volume of business and their credit standing. Some companies don't disperse that information freely.

Nail: Materials suppliers usually extend their regular customers (your contractor) a line of credit. When the economy is tight and work is scarce, some contractors don't or can't pay the company that supplied their materials. You can end up paying the contractor full contract price for your new roof, then find out a few months later that the materials supplier has placed a lien against your home for the cost of the materials. Find out who will be supplying the materials for your roof and give them a call. See what your contractor's track record is with them. It's still not a guarantee, but your call may save you a real shock.

Below is an example of a Lien Waiver Form, which homeowners can add to the contract.

5. Does your contractor have the company name and phone number painted on his trucks and other equipment? Beware of the "gypsy" company with no name and no phone number on anything.

Nail: There is a nasty scam which can (and does) happen. A highly regarded contractor will replace several roofs in a neighborhood. The contractor's equipment is all painted distinctive colors and his name is on everything. You get a knock on your door and a pickup truck is in your driveway with the same distinctive colors you have seen all over your neighborhood. The man says, "Hi, We're working here in your neighborhood. Would you like us to replace your roof while we're here?" You think he's from the highly regarded company. You haggle and hash out the details and agree to the terms of a contract. When you start to write the deposit check you find out it's not the same company at all. He is working on his own. It's surprising how many people, having made a verbal commitment, will go ahead and sign that check even though they know they have already been misled. If you push the man about tricking you, you will probably find he has done a home within a five-mile radius (which he considers to be your neighborhood). He will point

LIEN WAIVER FORM

The contractor/materialman (vendor) referenced below hereby acknowledges payment for all services and materials provided to _____ (customer), for the property located at _____, up to and including this date _____.

Vendor is known as _____, and has provided the following services and/or materials:

_____

_____

_____

_____

For these services and/or materials, vendor has received in payment, the sum of $_____.

With these payments, vendor releases any lien interest he may hold against the above-referenced property and its owner as of this date.

_____   _____
Customer   Vendor

out that there is no sign on his truck and he never said who he worked for. He is sorry you assumed he was working for the other contractor. He will give you a nice speech about his absolute honesty. All you have lost is your time and patience. Cut your losses and get rid of him.

6. Well-maintained and clean or new equipment speaks well for a contractor, but beware if he doesn't have his name and number on anything.

7. Signs held on with magnets don't impress me. The contractor can pop that sign off tonight and be a different type of contractor with a different name tomorrow.

8. Call the number the contractor gives you. Too many people call after they have already lost their deposit and find that there is no such number or the phone has been disconnected.

9. Get the license number and a description of the contractor's vehicle.

10. The contractor should furnish you with references including names, addresses, and phone numbers of previous customers. Don't just call them; take the time to go look at the work done on a few of the homes. A satisfied customer will be pleased to talk to you or show you the outstanding work he received.

11. How carefully did the contractor check and measure your roof? Did he just glance at it from the ground and give a figure?

12. Is the contract proposal on a company form with company stationery in a company envelope or is it on a "buy-em-by-the-box" form you can pick up at any office supply place?

13. Is the work spelled out in detail including brand, quality, and color of shingles? Does the contract specify the type of flashing, plumbing vents, and aluminum trim to be used? Are the type and size of fasteners listed? Is all of the work to be done included in the contract?

Don't do what too many others before you have done. Don't sign a contract with, or give a check to, someone who gives you a contract as vague as the following:

Mr. and Mrs. Reader
13 Sucker Punch Place
Dumfries, VA 22000

Shingle roof	$ 3,000.00
DEPOSIT DUE NOW	$ 1,500.00

Signed,

ELABORATELY ILLEGIBLE

14. Go by the contractor's place of business. If he is working from his home, don't hold that against him. It helps him hold down his overhead and give you a more competitive price. How is his building or home maintained?

15. Check with his insurance company and make sure he has liability insurance sufficient to cover possible damage to your home.

16. If the contractor does not carry worker's compensation insurance and he or one of his men is injured on your property, then the injured worker places his claim against your homeowner's insurance and you personally! It's not legal, but many roofing contractors "run bare" because of the high cost of worker's compensation. If one of these ends up being your contractor, you are "running bare" with him.

If the contractor does have worker's compensation insurance, he is precluded by law from filing a suit against you or your homeowner's insurance.

Nail: When a contractor lies to you about having insurance, he is trying to nail you by forcing you to take a risk you aren't aware of. You may be lucky and get by without damage to your home and with no personal injury accidents. You were still nailed.

Incidentally, if he succeeds in nailing you, his costs are much lower than the insured contractors he is competing with, and he, and you, have nailed the quality, insured roofing businesses in your area.

17. If you are negotiating with a contractor and you get a bad feeling about the whole deal, trust your instincts. Keep looking!

18. At this point the negotiations can still be blown. Let's say that you have done all of the above things and your contractor has passed with flying colors. You are dealing with a top contractor who now knows that you know it. Top contractors do mainly referral work, and they have customers waiting. Now it is your job to be the kind of customer this contractor wants to deal with. Listen to his recommendations regarding the work to be done. Remember, a top contractor will walk away from a potential customer he believes will be impossible or unreliable.

19. Now you can sign the contract proposal. Keep a copy signed by both you and the contractor.

20. You can expect to pay up to one-half the cost of the project as a deposit to the contractor. This will cover his purchase of materials and other startup costs. Your deposit assures the contractor that his customer won't do what an ex-friend of mine did to me; arbitrarily cancel the job the day the special-order materials are to be delivered.

21. A good contractor will have the materials delivered and (if possible) stocked up on your roof a few days before he has the work scheduled. If an independent supplier makes the delivery, expect the contractor to come by and make sure all the materials are in place.

22. Don't expect the contractor to negotiate on items like the requirement for a deposit or payment on completion (POC). He may have started out contracting without requiring a deposit and run into a customer who took delivery of the materials and then wouldn't let him "trespass" to do the job or retrieve his materials. He was then out the price of the materials and the job; his only recourse was through the courts. Your contractor can be a "prince among men," but he is probably also a battle-scarred veteran of the contracting game.

On pages 17–19 is a sample of the contract I developed and used. The thoroughness of this contract proposal swayed some prospective customers to contract with my firm for their roofing project. The contract is detailed, perhaps too detailed, but there was never a dispute or even a question about what work was to be done. We had a great relationship with our cus-tomers. The second year we were in business all but one of our customers came to us as referrals. A comprehensive contract followed by our workmanship guaranteed the referrals.

COMPANY INFORMATION SHEET

Established 19—
A Virginia Corporation
Virginia Class "A" Contractor, License Number 0123-45678
Fairfax County Business/Professional License Number 98765-43-21
Insurer: Great Plains Insurance
            1 Plains Way
            Richmond, VA 23220
Agent: Mr. I.M. Derevoc   Phone: (804) 555-6543
Liability coverage: YES
Worker's compensation coverage: YES
Insurance on all vehicles: YES
A member of Dun & Bradstreet: DUNS Number 567-890/12

SAMPLE CONTRACT

VISTA ENTERPRISES, INC.   May 1, 1993
5500 Summit Street
Centreville, VA 22020-2032
(703) 631-3366

Elizabeth Tomlinson
110 Prosperity Lane
Fairfax, VA 22033

Home Phone: (703) 555-1234   Work Phone: (202) 555-4321

Subject: ROOFING OF RESIDENCE
110 Prosperity Lane

Dear Ms. Tomlinson:

This letter is a formal contract proposal for the reroofing of your home. The work will consist of the items indicated on the "Contract Work Items" section of the contract attached to this letter.

The total price of the contract is $_____.
A deposit of $_____ is required.
The remaining $_____ is payable on completion.

This proposal will remain in effect until May 15, 1993, and can be formalized by your signature and the receipt of the deposit check.

COLOR OF SHINGLES   GRAY FROST

_____   _____

Elizabeth Tomlinson   John W. Chiles, Jr.
                                               President

Enclosures:
Information Sheet about Vista Enterprises, Inc.
Contract Work Items
Shingle Color — Examples of Gray Frost are noted on list of recent customers.

CONTRACT WORK ITEMS

TEAR-OFF:
Tear off the existing roof to the plywood/plank sheathing.
Install ice shield along the eaves.
Install No. 15 felt.

BASIC ROOF:

Install Certainteed Glassguard, self-sealing, fiberglass/asphalt shingles with a Class A U.L. fire resistance rating. These shingles carry a twenty (20) year manufacturer's warranty.

Use one and one-half (1½) inch galvanized roofing nails.

Valleys will be shingled as indicated on the options list.

Plumbing vent pipes will be flashed using aluminum-based neoprene collars.

Clean and tighten gutters.

Police yard and flower borders thoroughly.

PRICE OF BASIC ROOF: $_____

PRICE OF CHOSEN OPTIONS: $_____

TOTAL CONTRACT PRICE: $_____

OPTIONAL CONTRACT WORK ITEMS

Install Certainteed Glassguard 25 self-sealing fiberglass/asphalt shingles with a Class A U.L. fire resistance rating (twenty-five [25] year manufacturer's warranty.) $_____

Install Certainteed "Independence Shangles," self-sealing fiberglass/asphalt shingles with a Class A U.L. fire resistance rating. These are dimensional shingles that simulate the surface texture of a cedar shake roof. The shingles have a thirty (30) year manufacturer's warranty. $_____

All-lead plumbing pipe vent flashing $_____

Aluminum step flashing $_____

Aluminum counter (or skirt) flashing $_____

Reuse existing aluminum valleys $_____

Valleys constructed of twenty-four (24) inch wide, 27 gauge aluminum over an underlayment of No. 30 felt $_____

Valleys constructed of double woven shingles over an underlayment of No. 30 asphalt felt $_____

Forty (40) inch wide modified bituminous sheet used as an underlayment for the valley $_____

Flash the brick/stone chimney $_____

Caulk cracks in the chimney cap $_____

Install a new concrete chimney cap $_____

Wire brush, paint, and caulk the metal chimney $_____

Wire brush, paint, and caulk the existing roof vents $_____

Replace the existing roof vents with new aluminum pot vents
(Type: _____) $_____

Cut sheathing and install brand new vents (Number: _____ Type: _____)	$_____
Install _____ feet of aluminum ridge vent	$_____
Install all-aluminum power ventilator (Brand: _____ Model: _____)	$_____
Aluminum flashing around skylight	$_____
TV antenna	$_____
Solar panel	$_____
Cupola repair and replacement	$_____
Plywood/plank sheathing replacement	$_____
Aluminum drip edge (above the gutter)	$_____
Aluminum roof edge (up the rakes)	$_____
Install 5-inch "K" seamless gutter (Color: _____)	$_____
Install 6-inch "K" seamless gutter (Color: _____)	$_____
Downspouts	$_____
Other	$_____
TOTAL PRICE OF OPTIONS	$_____

REFERENCE AND SHINGLE TYPE/COLOR LIST

The list included the following for each home we reroofed during the previous two years:

Owner's Name   Home Phone
Street Address
City, State, Zip Code

Manufacturer, type, and color of shingle

(I never had anyone request that they not be included in this list of previous customers. I didn't realize this until I sat down to write this book.)

A contract is a legally binding document about something as nebulous as mutual trust. A contract obligates both parties to perform. It takes a good contractor and a good customer to give top quality results.

I caught a glimpse of the shingles hanging over the front rake of this garage and stopped to talk to the homeowner working on his roof. He was having problems and agreed to let me take pictures if I would show him how to do the rest of his roof.

He was starting his base course shingles flush with the face of the rake board and that did not provide a continuous protective overhang over the rake.

When I saw the rear roof of his garage, I realized that he wasn't initially aware of the 6-inch offset between alternating courses.

A view of the top of the rear rake.

When I met him, he had already shingled partway up his front garage roof. He had looked at other roofs and figured out the 6-inch offset. He was still starting his base course even with the face of the rake board. He just measured 6 inches over to set his offset course. He was setting his horizontal courses by measuring up 5 inches from the lower edge of the tab of the shingle below. He didn't have a chalk line.

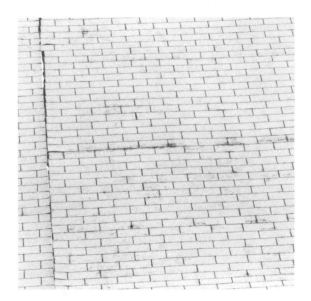

Two-thirds of the front of the main house is this 12/12 roof. It is a major architectural component. He used 2 x 4 toe boards instead of roofing scaffold, and now he has a pure white roof heavily lined all the way across with scuff marks. This man's hard work with expensive materials should never have turned out this badly. By the time he got to his main roof, I know he had the knowledge to do a quality job. Sometime in the future, some poor real estate agent will grope for the words to tell him nicely how much his botched roofing effort has devalued his home.

2.
Basics

Whether you will contract or do it yourself, there are specific decisions that need to be made. Color is extremely important not only aesthetically, but also as a matter of energy efficiency and durability. Installation of additional ventilation including power ventilators is important for comfort and energy consumption. Tool and clothing tips are important to the do-it-yourselfer. Read the following sections and combine the facts (and opinions) with your personal preferences and tastes, so you can decide which best suit you and your home.

COLOR

The question that almost always takes the most time and thought is color. The color of your roof not only complements the exterior of your home. It also affects the comfort of the interior, future energy consumption, and the life of the roof itself.

Manufacturers distribute small sample pads of their various colors of shingles. Most shingles today have a random patterning. Manufacturers try to show you the prevalent colors in their samples. If you have narrowed the color choice down, your roofing supplier also has larger sample boards (approximately 2' x 2½'). These are sections of the roof as it will look after it is laid. Holding a sample board up against the various exterior features of your home will give you a good idea of how the colors will work together. The small samples and larger boards can be borrowed from your supplier. If you are dealing with a contractor, he may be able to refer

you to a home with a color combination similar to or exactly matching the one you are considering.

Manufacturers have set up test roofs with their various colors and types of shingles. Their tests show that a *black* roof absorbs three times more BTU's from the sun than a white roof! (A BTU, British Thermal Unit, is the amount of heat required to raise the temperature of one pound of water one degree F.) In the summer, black roofs can actually burn your hands, scorch your rear end, and blister your feet through your jogging shoes. The darker the roof, the more heat you are soaking into your attic. A dark roof will throw so much additional heat into your attic that your existing ventilation can't get rid of it. Then the heat spreads through your attic insulation and down into your living space. This either makes you more uncomfortable or makes your air conditioning work much harder. Now we are talking additional costs all summer, every summer, as long as that dark roof exists.

If a black roof is the only thing you would dream of, then make sure you also put in plenty of ventilation. Ridge vents can be installed along the peak of your roof, or domed vents (pot vents) can be installed in the back (or somewhere not noticeable). Power ventilators with automatic thermostats can be installed. Small diameter metal vents can be installed in the soffits (the horizontal board or plywood under the eaves).

Assuming you have an insulated attic, a black roof will not help heat your home in the winter. It is true that even in winter the black roof continues to absorb more heat than the white one. However, if your attic is properly ventilated, the cold winter winds will carry the heat in your attic away before it can help heat your living space.

Don't even think about blocking the attic ventilation in the winter. This would make your attic "sweat" with condensation and trapped moisture from your living space. The sickening smell of rot would soon prevail.

Why do so many new homes have black or dark roofs? When a roof gets hot, the asphalt sealing the grit to the shingle gets more viscous. When the asphalt holding the grit gets soft, it is very easy to scar the roof. On a white roof, you may either have to work on a shaded section of it or get off it completely during the hottest part of the day. If you scar it, you have to replace the damaged section, or you will see it from the ground from then on. If a black asphalt scar happens on a black or dark roof, you can't see it.

Nail: I have seen contractors use toe boards (2 x 4's nailed into the roof to enable the crew to stand on steep roofs). They nail them right through the shingles on new townhouses. The nail holes through the new roof were either sealed up with roofing cement or worse, not sealed at all. The crew's shoes had rubbed most of the grit off the new shingles above the board. It was a black roof and you couldn't see it from the ground, so they left it that way.

Color also affects the life of the roof. Here are some important factors in the aging of your shingle roof:

1. The black roof, which gets much hotter than the white roof during the day, drops down to the same temperature as the white roof at night. This means the black roof is experiencing a much wider range of expansion and contraction.

2. The black roof also absorbs many of the ultraviolet rays that the white roof is reflecting. Ultraviolet rays age and harden the asphalt

component of the shingles. This tends to make the shingle more brittle.

3. Heat tends to bake off the oil in the asphalt component, making the shingle more brittle.

It has been my observation in a subdivision of 1,850 homes that black roofs needed replacement two to four years earlier than white roofs. Variations were caused by shade trees, owner-added attic ventilation, and orientation to the sun.

On one split foyer home (basically a two story) in the middle of the summer, we overlaid a black roof with a new white one. The four days after we finished the work were the hottest, most humid four days of the summer. Our amazed customer called me at the end of those four days and said that during those four days, his upstairs bedrooms were cooler and his air conditioner had run less than it had since spring.

On some homes, a black roof does looks more elegant than any other imaginable color. Sometimes a black roof keeps the roof itself from drawing attention from more important architectural elements of the home. I confess that I installed two black roofs (under protest). I only did so after both customers agreed to adding considerably more ventilation.

You may be faced with deciding between a dark brown or a light brown shingle, and you know you would be equally pleased with the appearance of either one. You are now armed with information. My advice is always to go with the lighter-colored roof.

SELF-SEALING SHINGLES

Most modern shingles are self-sealing. These individual shingles have an asphalt strip across the width of the shingle aproximately 5 to 6 inches from the top. This strip is there to seal down the lower, exposed edges of the shingles which will be laid in the next course up the roof. When the sun hits your new roof, the self-sealing asphalt strip gets gummy from the heat, and the asphalt works its way into the back of the shingle resting on top of it. A day of hot sun on a black roof and it is sealed down. A white roof will take three days to a week or more to seal.

The old thick organic shingles didn't have this self-sealing feature. Those old design shingles depended on the rigidity of the shingles to keep the wind from lifting them up and tearing sections from the roof. I can't think of any case where it is preferable to use a shingle that is not self-sealing.

Nail: A contractor can get still get cheaper shingles that are not self-sealing. If you stressed low bid too much and didn't know about this self-sealing feature, guess what you would get.

Shingles come in a paperbound bundle. What keeps the bundle from sealing itself into one solid mass? Self-sealing shingles have a strip of plastic or tape across the back. This protective tape is located on the back, just above the self-sealing asphalt strip on the top of the next shingle down in the bundle. This keeps the self-sealing strip from melting into the back of the shingle above it in the bundle. This protective tape across the back is left attached in place when the shingle is laid. It has already done its job of keeping the shingles separated in the bundle. A few of the protective tapes will come partially loose as you pull a shingle from the bundle. Just pull that occasional tape free and dispose of it.

A neighbor of mine roofed his own home before I knew him, and naturally we got into a discussion about roofing one day. He said one of the things that aggravated him the most was pulling those tapes off the back of each shingle before he nailed it in place. That was when I first thought of writing this book.

PLUMBING VENTS AND OTHER FITTINGS

Your plumbing vents will need *vent flashing* to seal them up. Most roofers in our area use a neoprene (black rubber) collar with a galvanized metal base. These rubber collars crack or disintegrate before the roof has served its useful life. If the neoprene collar is your only choice, you will probably need to seal around it with roofing cement or caulk, starting

when it is aproximately ten years old. Don't put a bunch of plastic stuff up on your roof. You can buy plastic ridge vents, power ventilators with plastic domes, and plastic vents. Plastic fittings almost always crack or crumble prior to the end of the useful life of the roof (ultraviolet rays at work again). Aluminum generally costs more, but it's worth the extra cost for its durability.

NAILS VS. STAPLES

Talk about a controversy among roofers! Contractors are moving rapidly to pneumatic (air) tools, whether the tools are pneumatic nailers (nail guns) or staplers. The arguments over using nails or staples are endless. Most of the roofers in our area use pneumatic (air) staplers. Unfortunately, all of the roofs that I have seen damaged by forty- to fifty-mile an hour winds have also been stapled. OK, you know which side I'm on now.

There are some things about staples that you need to know. Staples are cheaper and so are staple guns. Staple guns are, arguably, quicker, lighter, and have less kick to them than pnuematic nailers. I observed that my crew would much rather tear off an old roof that was stapled. The prospect of a tear-off normally brought groans, but we started referring to the tear-off of a stapled roof as "unzipping" it.

Air guns will take up to $1^1/_2$-inch roofing nails, which come in a coil. If you contract your roof, try to get a contractor who uses nail guns and insist on the $1^1/_2$-inch nails. The $1^1/_2$-inch nails go in straighter than the shorter nails, and a hurricane may tear some of your shingles, but those nails aren't going to pull loose. With nails, you won't have a fifty-mile-per-hour wind rip away a 10' x 10' section of your roof.

If you are going to do it yourself, don't buy or rent a compressor or nail gun. It's not worth the money. For one thing, as with most do-it-yourself projects, the job is going to take longer than you think. Working on the sloped surface is going to throw an unaccustomed strain on you, and you will find that you can't work as long each day as you think you can. Take the slower, more tedious route of hand nailing your shingles. The 1-inch galvanized nail is the standard, but the 1¼-inch or, better yet, the 1½-inch will diminish the probability of smashing your fingers. Hold the shaft of the nail between your forefinger and middle finger *with your fingernails down against the shingle*. That way, if you miss the nail, your hammer hits the fleshy part of your finger and you aren't as likely to smash your fingernail.

TOOLS

You need basic tools to do roofing. They include:

1. Hammer: A basic claw hammer will do the job. Although a straight claw hammer with the flatter, more swept back claws is handier for some jobs, don't buy a new hammer for the difference.

2. Tape: A 25-foot retracting steel tape is the best all around for this work.

3. Knife: A Hyde® knife or another brand of utility knife with a similar crook in the handle is preferable for roofing work. The crook raises your knuckles up and away from the work. A straight-handled knife will end up letting you rake your knuckles across the shingles. The price of the Hyde® knife (usually $6.00 to $7.00) is worth the saving of the "hide" on your knuckles.

4. Knife blades: You can buy straight blades and hook blades for your Hyde® knife. If your knife doesn't come with blades, buy a small pack (three to five blades per pack) of each type.

5. Hammer hook and nail pouch: If you don't already have them, invest in a hammer hook and nail pouch for your belt. Don't buy the swinging hammer hook. Carpenters like the swinging type, but when you sit down on the roof, your hammer handle will hit and push up and out of the swinging hook. Watch your hammer slide down and off the roof! Buy the hook riveted to the leather backing your belt goes through.

6. Chalk box: You are going to need a chalk line. A 50-foot line is plenty. If you use a longer line, it sags down the roof, and your courses will be bowed down as you sight across them. A good heavy-duty chalk box is the best. Don't go for one of these fancy "speed" boxes that have gears inside to help you reel in the line faster. I never saw one on the job that wasn't broken. Use blue chalk. The other colors such as red or yellow are permanent. If you drop your box or spill chalk over your new shingles, the color will stay there for a couple of years. Hosing and sweeping it down won't remove it.

These are all basic low cost tools, and you may already have them. In addition, you may need a hacksaw for the plumbing vents, a circular saw, and sheet metal shears.

CLOTHING

The sun is your enemy up on the roof. You are going to be getting rays from above and also have them reflected back at you. Stay covered up, including a cap or brimmed hat. Long sleeves are a good idea if it isn't too hot.

Even the best work gloves are hot and wear out fast as you handle shingles. (Think of shingles as large, rigid sheets of extremely coarse sandpaper.) Instead of using gloves, get a roll of 1-inch adhesive tape and wrap your fingers loosely with it. Don't wrap around your knuckles; you'll need the flexibility. There is no need to cover your palms; your fingertips get most of the wear and tear. Get the kind of tape that won't pull your skin off. I learned this trick from a former football player; it will sure save your hands, and it's much cheaper than gloves.

Shoes are a critical item. A pair of old jogging shoes is ideal. Worn soles won't cut into the surface of the new shingles as readily as new soles will. Don't wear work boots and especially *don't wear anything with leather soles* on the roof. You want to stick up there like a fly; you need maximum flexibility for your ankles and maximum traction on the bottom of your feet.

Some men have found boots protected them from nails. Boots don't give you the traction and flexibility your feet need. Don't worry about getting a nail in your foot. Keep the roof and job site cleaned up instead of worrying. Besides, you'll develop "smart feet." You will feel the pressure of the nail through the sole of your shoe and will freeze before it gets all the way through. Even if a nail should get you, it's better to be in a little pain on the roof and accept the tetanus booster than be in a whole lot of pain sprawled on the ground.

The difference in blend numbers is glaring on the front of this townhouse.

The roofer did take the trouble to take his odd-numbered bundles to the back of this two-story home.

3.
Roof Computations for an Overlay

If you are doing the work yourself, you need to order your materials. There is only one way to find out what you need and that's to get up on the roof and measure. But first, if there has been a problem with leaking, go up in your attic and check for rotten sheathing (plywood or planks) and rotten rafters. If an area of the roof is bad, you'll see it from the attic and won't get any surprises (like falling through) when you walk on the roof.

LADDER

The "feet" on your ladder should be anchored firmly in the grass at a distance from the wall equalling approximately one-fourth the height to be climbed. Don't set up on a concrete sidewalk or patio if you can help it. A smooth, hard surface increases the chance that the feet will kick out at the worst possible time.

If you are using an extension ladder, pull the slack out of your rope through the pulley. Tie the rope to adjoining rungs on both sections of the ladder. If the rope used to raise the ladder is gone, tie two adjoining rungs on each section together securely. You don't want the catches to fail and the extension to slide down with you on the ladder.

If your home has gutters, set your ladder up in the rear or on some other less visible section of gutter. If the ladder scrapes or damages the gutter, it won't show as much. Center the ladder over a spike and ferrule (a long nail and a spacer sleeve around the nail). This is the strongest point on your gutter. Make sure three or four rungs extend above the gutter. The additional length of ladder will give you something to hold on to when swinging from the roof back onto the ladder. On the first trip up, tie the ladder securely to the spike and ferrule to keep the ladder from shifting or blowing down. Hook a bungee cord over the spike and ferrule, loop it around your ladder, and hook the far end of the bungee back to the same spike and ferrule. Your ladder won't go anywhere.

If your home doesn't have a gutter, tie the top of the ladder off to a chimney, plumbing vent, etc. If nothing else is available, you can drive a couple of strong nails partway into the fascia (the edge above the gutters) and tie off to the bent-over nails.

PITCH OR SLOPE

Shingle roofs slope up from the horizontal. Roofers don't talk about this slope as a vertical angle. They don't say, for instance, that the roof is at a 45° angle; they'll say the roof is a 12/12. The slope is measured by the rise (vertical distance) over the run (horizontal distance). Let's say that in a 1-foot (12-inch) horizontal run, your roof rises 4 inches. This is referred to as a 4/12 roof. To find out what your slope is, hold the end of a bubble level against your roof surface and measure 12 inches. Take a vertical measurement down from the 12-inch point on your level to the surface of the shingles. For example, if your roof drops down (which is the same as rising) 5 inches in 12 horizontal inches, you have a 5/12 roof.

If your roof is steeper than 6/12, you will be smart to let a professional roofer do it for you. Professionals refer to up to an 8/12 as "walkable," but part of that is macho bragging. True, you can walk and work on such a roof without scaffolding. However, you can fall near the peak of a 7/12 roof, claw and grab as you slide down, and still pick up speed all the way to the ground.

ROT

Walk around on your roof, checking its general condition. The 2 x 4, 2 x 6, or heavier supporting rafters underneath your plywood or plank sheathing are probably set on either the standard 16-inch centers or 24-inch centers (measured center to center of the rafters). Walk the roof, checking for areas that give or feel soft underfoot. If you feel something giving too much, get your feet over the rafters quickly. If the sheathing gives way completely, you could find yourself back in your bedroom in a hurry.

BASIC ROOF LAYOUTS

There are five basic types of roofs. They are:

1. Lean-to: The lean-to roof is a sloped roof that rises up to tie into a wall. You see this type on screen porches, carports, etc.
2. Saddle: The saddle roof is a roof that rises straight up the front to a ridge line, then drops straight down the back. The measurement of both legs of roof from the ridge is usually the same. The gables at the end of a saddle roof rise straight up from the end walls to a point at the ridge. Roofers call a saddle roof an *up-and-over*.
3. Hip: The main part of the hip roof is an up-and-over but instead of having gables, a triangular section of roof slopes up from the top of the

end walls to the ridge of the roof. Some people call these "Dutch" roofs.

4. Gambrel: This is a roof having two distinctly different sloping sections on both the front and the back legs of the roof. The lower roof sections have a steeper slope, and the upper sections of the roof have a flatter slope. The gables on each end are vertical. Some later model townhouses have the gambrel roof with dormer windows projecting from the lower, steeper sloped section of the roof.
5. Mansard: A roof having two slopes on both front and back with the lower slope steeper than the upper one.

MEASUREMENTS

Make a "bird's-eye" sketch of your roof. Show the complete outline of the roof at the fascia. I have basically drawn my figures to scale (1" = 10'). Your roof is sloped, but draw it as if it were flat. Show all valleys, ridges, plumbing vents, pot vents, metal chimneys, brick chimneys, skylights, power ventilators, etc. Don't leave anything out of the sketch, or you might forget to order something. (Small items are not pictured in Figures 1-3 in order to simplify things.)

If you have a "straight up-and-over," you are in luck. Your roof goes up the front and down the back with no additional wings, porches, or dormers to play with. This is the simplest kind of roof. Even if your roof is this simple style, you should make a sketch of the roof showing everything on it.

Let's say your roof is like Figure 1. The length of your roof along the peak or "caps" is 50 feet. Your measurement up the slope from the fascia (at the gutters) to the peak is 20 feet.

To figure the squares of shingles (10' x 10' of roofing) you will need, just add an additional foot to the measurement up the slope to cover the *starter course* along the fascia and the *cap pieces*. For the normal up-and-over roof, the measurement down the back slope (the other half of the roof) will also be 20 feet. Don't ass/u/me this; measure it. Add an

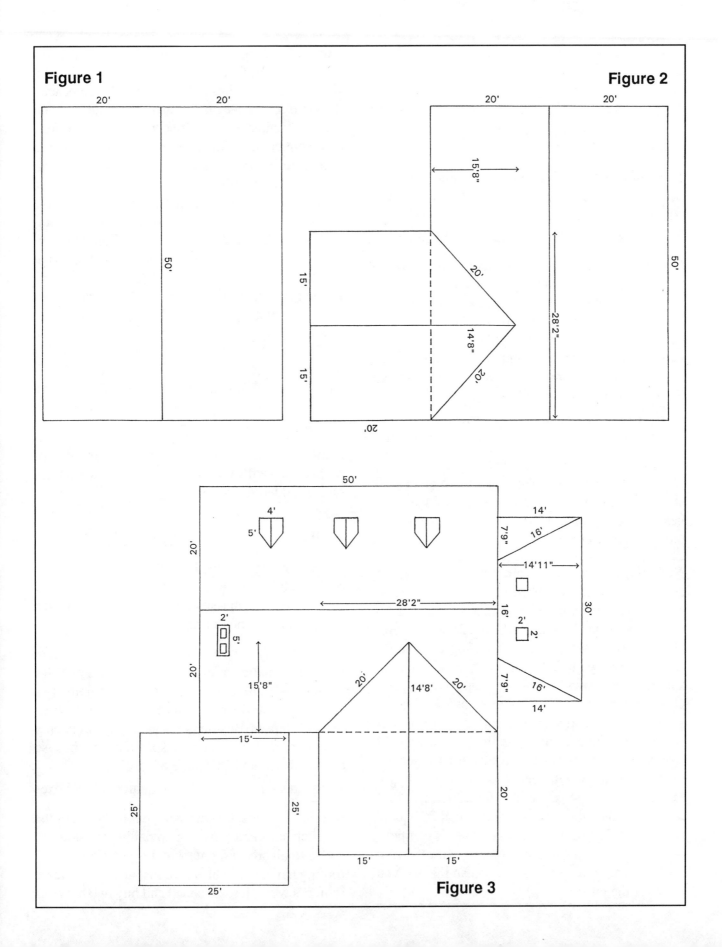

Figure 1

Figure 2

Figure 3

additional foot to the measurement of the rear slope too. The math is:

50' x (20 + 1') + 50' x (20 + 1') =
2 x 50' x (21') = 2,100 sq. ft.
2,100 sq. ft./100 sq. ft. per roofing square = 21 squares

The standard three tab shingle comes one-third of a square to a bundle, or three bundles of shingles equals one square.

21 squares x 3 bundles per square = 63 bundles

OK, we're rolling, except that there will be a few mistakes, and some of the shingles will be damaged when you open the bundles. Let's order an extra square of shingles just to be on the safe side. (You can always take your receipt and turn in any unbroken, undamaged bundles and get your money refunded, minus probably a 15 percent handling fee.) So now what do we order?

21 squares + 1 extra square = 22 squares or
22 squares x 3 bundles per square = 66 bundles

Note: If you are ordering *dimensional shingles*, which more or less simulate the texture of a cedar shake roof, you will find that these shingles are heavier than standard three tab shingles and are packaged four bundles to a square.

Nail: If you haven't measured and computed the area of your roof yourself, your roofing contractor can get you. He can tell you that you have a 30 square roof when in truth you have a 25 square roof. He will give you his price for roofing 25 squares, while telling you that your roof is actually 30 squares. He knows you might want to do a quick check on him and may find another roofer willing to give you a rough price over the phone. You use your first roofer's figures and report that you have a 4/12 straight up-and-over that is 30 squares. The second roofer's quote over the phone is automatically going to be twenty percent higher for a 30 square roof than his quote would be for your actual 25 square roof. Your first roofer has edged out his competition, and you feel good contracting with the first roofer because his price is lower than the one you got over the phone.

You ordered an extra square because you don't want to go back to your supplier for additional shingles. The bundles of shingles may all come with the same color code on the end of the bundles, "Star White" for instance, but one production run may come out cream white and another production run will come out chalk white. The production run number (or blend number) is on the end of the bundle below the color code number, or it may be on the opposite end of the bundle from the color code number, depending on the manufacturer. Tell your supplier you want not only the color code numbers but also the blend numbers on all the bundles to match. If you get mixed blend numbers delivered to your roof, you won't be able to tell the difference as you're laying the shingles, but it's very possible you'll be able to see it from the ground. To put this in perspective, ask someone who knits what it would be like to use different dye lots of yarn in one sweater. Then brace yourself for the answer.

Of course, if your supplier slips you a few odd-numbered bundles, and you also have a section of roof nobody will ever see, just use your odd shingles on the hidden section.

NAILS

If you order a 50-pound box of 1½-inch galvanized roofing nails, you may have enough for your roof. If your house is massive, you can always go back and get another box of nails.

If you already have a compressor and you want to buy a pneumatic nailer (also called a nail gun) for $400 to $450, here are some figures pertaining to a specific brand. Bostitch nails are made for Bostitch guns. A box of Bostitch coil nails will do twenty squares of roofing. There are sixty coils to a box, so each coil will nail ⅓ square of roofing.

22 squares x 1 box of nails/20 squares = 1.10 boxes

In a case like this, I recommend buying a box of the Bostitch nails and finishing up nailing the shingles with a small box of regular hand-dipped galvanized roofing nails. If you have, for instance, a 32 square roof, it's worth it to go ahead and buy two boxes of the Bostitch coil nails. (Suppliers won't accept the return of a partial box of coil nails.)

Now let us suppose your house has a wing or is L-shaped. (See Figure 2.) A triangular area of your main roof in the rear is now nonexistent: it is covered over by the two triangles that tie your wing into the main roof. This makes your math more interesting. The main house has the same measurements as Figure 1.

Main House: 2 x 50 x (20 + 1) = 2,100 sq. ft.

The roof area of the wing is broken down into two rectangles and two triangles. (The area of a triangle equals one-half the base times the height: A=½BH).

Wing: [Area of two rectangles] + [area of two triangles]
[2 x 20 x (15 + 1)] + [2 x ½(15 + 1) x (14'8")] =
(Convert inches to hundredths of a foot by dividing by 12"/ft.
8"/12" per foot = 0.66')
[640] + [2 x 8 x 14.66] =
[640] + [234.56] = 874.56 = 875 sq. ft.

Now, subtract the triangular area of the rear roof of the main house that is nonexistent due to the triangles of the wing tying in above. You can sight down from the ridge and "guesstimate" the base of the triangle that has been covered over. You can also measure up the roof "guesstimating" the height of the covered triangle.

[½(28'2") x (15'8")] =
[½(28.17) x (15.66)] = 220.57 = 221 sq. ft.

So how many squares of shingles do we order?

Main house = 2,100
Wing   = 875
Minus area = < 221>
Total   = 2,754 sq. ft.

Now after all this exactness, we need a fudge factor. Notice you have two 20-foot valleys where the wing roof joins the main roof. No matter how you finish a valley, it's going to gobble up a lot of shingles. With valleys that long and with the probability of some damage in shipment or of making one or two errors, you should order 2 extra squares of shingles.

2,754 sq. ft./100 sq. ft./roofing square = 27.6 squares

You will need: 27.6 squares + 2 squares = 29.6 squares

Remember that roofing comes in ⅓ square bundles. 29.6 squares x 3 bundles/square = 88.8 = 89 bundles

Now let's suppose you own the two-story home in Figure 3. In addition to the rear wing we saw in Figure 2, you have three dormer roofs on the front and a lean-to porch roof tying in partway up the left side of the back wall. There is a brick chimney. A sunroom with two skylights ties into a side wall.

First let's decide what we are *not* going to worry about. Let's figure the dormers as if they were three simple rectangles instead of two even smaller rectangles and two very small triangles:
[3 x (5 x 4)] = 60 sq. ft.

The two squares for the skylights are:
[2 x (2 x 2)] = 8 sq. ft.

The brick chimney is:
5 x 2 = 10 sq. ft.

Three dormers: <60> sq. ft.
Two skylights: < 8> sq. ft.
Chimney: <10> sq. ft.
Total   <78> sq. ft.

78 sq. ft. x 3 bundles/100 sq. ft. = 2.34 bundles

You already know that roofing the dormers — "cutting into" the walls of the dormers, skylights, and chimney — will take extra shingles. If you have small items such as skylights and chimneys on your roof, just figure the roof as if they weren't there. If you have valleys, figure your areas and add a couple of bundles rather than trying to figure everything out to the nearest hundredth of a square.

Here we calculate the whole thing:

Main House: [2 x 50 x (20 + 1)] =   2,100 sq. ft.
Rear Wing: [2 x 20 x (15 + 1)] + [2 x ½(15 + 1) x (14'8")] = 875 sq. ft.
Minus: [½(28'2") x (15'8")] = <221 sq. ft.>
Screened Porch: [25 x 25 + 1] = 650 sq. ft.
Sunroom: [two triangles + trapezoid]

(The area of a trapezoid equals one-half the sum of the two bases times the height. $A = \frac{1}{2}(B1 + B2) \times H$)

$[2 \times \frac{1}{2}(14) \times (7'9" + 1)] +$

$[\frac{1}{2}(30 + 16) \times (14'11" + 1)] =$

$[122.50] + [366.16] = 488.66 = 489$ sq. ft.

TOTAL 3,868 sq. ft.

3,868 sq. ft./100 sq. ft./square = 38.68 squares

Round it off to 39 squares = 39 squares

But remember, we added a couple of squares for the valleys on the back. We should figure another extra square for the dormers and skylights and chimney altogether. Since you are new at this and the roof is a bit complex, add another $\frac{2}{3}$ square.

39 sq. + 2 sq. + 1 sq. + $\frac{2}{3}$ sq. = $42\frac{2}{3}$ squares

$42\frac{2}{3} \times 3$ b/sq. = 128 bundles

This is not an exact science, but you will be surprised how close you'll come. The secret is always to figure to the high side, so you won't come up short at the end of the job.

If you or your roofing contractor have a partial bundle left over, stash it away somewhere reasonably cool and dry just in case a shingle fails or is torn in a high wind later on. If you have a couple of unopened bundles left over, return them along with your receipt to your supplier. Remember, you can get a percentage (usually 85%) of your purchase price back.

VENT COLLARS OR FLASHING

Your plumbing has vent pipes that go through the roof to get rid of methane and other gases from the sewer system. You can see the vent pipes through your roof above the kitchen, bathrooms, and laundry room. The plumber may have run one vent to adjoining bathrooms or say a laundry room and kitchen. These pipes stick up several inches to a foot above the roof.

The standard vent flashing is a neoprene (rubber) collar mounted on a galvanized metal base. These are the easiest to install; you just slide them down over your vent pipe and weave the base right in with your shingles. However, there are problems with these neoprene collars. Many of them fail before the

roof needs to be replaced. This allows water to penetrate the roof. You can use them if ten years from now you want to start getting on your roof occasionally and caulking around them with roofing cement (asphalt mastic) or silicone caulk.

Commercial all-lead flashing is the best option. It consists of a 12" x 12" flat lead base with a lead pipe soldered solidly around the hole in the center of the plate. This type is slightly more expensive and a little more time-consuming to install, but it will outlast your new roof without requiring any additional maintenance.

Plumbing vents are measured by their outside diameter. A 2-inch vent has a 2-inch outside diameter. Okay, let's suppose on your home you have a 2-inch vent and a 3-inch vent.

All-lead vent flashing:

1 2-inch collar = 2 x $ _____ = $ _____

1 3-inch collar = 1 x $ _____ = $ _____

VALLEYS

If your house has valleys, you need to pay special attention to them. Valleys are one of the most leak-prone locations on a roof.

If you have an aluminum or copper valley which has been trouble free, you will probably want to reuse it.

You can also "double weave" the valleys, alternating the shingles as you lay up both of the roofs on the main house and the wing at the same time. You can "double weave" or "California Cut" right over an existing copper or aluminum valley.

The California Cut means you lay the main roof and extend the shingles through the valley from one direction. You then run the shingles from the other roof through the valley also and cut the second, or top, layer of shingles off along a chalk line.

More details on these methods later. Reusing the existing metal valley is the most durable and best looking way to do your valleys. Double weaving may give a slightly rougher appearance than the California Cut, but it is considerably stronger and less likely to give trouble.

Nail: Most roofers will do your valley with a California Cut regardless of whether your metal valleys were giving trouble. It is the quickest and easiest thing for them to do. If it should leak or blow loose, your old metal valley is still in place and will work, assuming your roofer didn't put any additional holes in it.

If you plan to reuse your existing metal valleys, you will need a caulk gun and some tubes of roofing cement to seal the cut edges of the new shingles on both sides of the valley to the old shingles beneath them. *Don't drive a bunch of nails into new shingles in the area of the valley.* You will need a tube of roofing cement for approximately each 15 feet of cut edge you need to seal down.

In Figure 3 it would go like this:
Roofing cement 2 valleys x 20' valley x 2 sides =
   80'
80'/15' per tube = 5.33 tubes
= 6 tubes of roofing cement

From here on, roofing cement will be referred to as *mastic*.

STEP FLASHING

If your roof were like Figure 3, you would need step flashing for your chimney, skylights, and dormers. Step flashing would also be required where the triangular sections of the sunroom tie into the wall. The standard piece of step flashing is a piece of 5" x 7" aluminum, which is bent L-shaped to give 2-inch and 3-inch legs to the 7-inch piece. More about step flashing later, but now you need to know how much to order. For every 5-inch course of shingles, you need one piece of 7-inch step flashing. It is easiest to figure the number of pieces per foot:

12"/foot/5"/course = 2.4 courses/foot=
   2.4 pieces of step flashing/foot

We will tie in two 7'9" lengths of wall for the sunroom:

2 x 7.75' x 2.4 step fl./foot = 38 pieces

Step flashing along the sides of the chimney:
2 x 5' x 2.4 step fl./foot = 24 pieces

Step flashing along sides of skylights:
2 x (2' + 2') x 2.4 step fl./foot =
   20 pieces

Step flashing along the walls on the sides of the dormers:

3 dormers x 2 walls x 4.5' x 2.4 step fl./foot =
   64.80 pieces
Total = 147 pieces

If you are using aluminum, these pieces are pretty cheap. Go ahead and order 175 pieces of step flashing. Also order 1-inch aluminum siding nails. (The 1½-inch aluminum nails bend easily when you are driving through heavy stuff.) Aluminum nails are best, because you don't want to mix dissimilar metals. Electrolysis between dissimilar metals speeds up corrosion.

SKIRT FLASHING

The vertical leg of step flashing slides into place behind siding. On a brick wall, it goes behind the counter flashing or what I call "skirt" flashing. Skirt flashing is sealed along its top edge with caulk. To make skirt flashing, you will need a roll or "coil" of aluminum. Coils come in 50-foot lengths and are either 12 inches wide or 24 inches wide. The coils come in two different gauges (thicknesses). The heavier .027 gauge (.027 of 1 foot) costs a few dollars more a coil and is preferable to .019 gauge. These coils are also used to make metal valleys. If you are only going to use the aluminum along your walls or chimney, 12-inch coil will probably be sufficient, depending on how high the existing skirt flashing is and what old caulk will be left visible when you replace it. Retrace what we did on the step flashing and see how much coil you need. Add up the skirt flashing along porch walls, sides of chimney, sides of skylights, triangle roofs on sunroom, and sides of dormers.

15 + [2(7'9") + 16] + [3(4) + 6(3)] = 76.5'

In this case, you would round up to 100 feet because the coils are 50 feet long. You also have 2 20-foot valleys. Consider purchasing three 20-inch wide coils (150 feet).

Having done your own computations, you go to the roofing supplier with a complete list of materials. This tells the roofing supplier something. He knows that you know what you are talking about, and he will treat you differently than he will a homeowner who obviously knows nothing about what he is doing. Remember that he doesn't know you; he's going to quote prices in the higher range. Let him know you are shopping with a couple of suppliers and then do so. You are doing the hard work anyway, and serious bargaining can save you hundreds of dollars.

When you pick your supplier, he will want to be paid in advance. There is certain to be a delay even if all the material is right in his warehouse. He will want your check to clear before he delivers anything to your roof. A cashier's check can speed up this process. This is a little inconvenient, but the supplier is a battle-scarred veteran too.

Suppliers may refuse to deliver to your roof, or "load the roof," if the pitch is over a 6/12. There are two reasons for this refusal. The first is the danger of bundles sliding off the roof, doing damage to gutters and shrubs, and the liability if they should fall and hit someone on the ground, causing serious injury or even death. The second consideration is the increased risk of falls to his men.

Your supplier won't deliver to your roof if cars are blocking your driveway, if he has to back up onto your grass, or if tree limbs or power lines prevent him from swinging his conveyor boom around from its cradle over the cab of the truck.

There are proper and improper ways to load the roof. The shingles should be in smaller stacks, spaced out on your roof. Each stack should contain no more than sixteen bundles (1,200 lbs.) Putting more weight than that in one spot on your roof is courting disaster. There is also a proper way to stack the shingles. The delivery person should lay one bundle just below and horizontal to the caps. The

Proper stacking of shingles on a steep roof.

next bundle should rest with one edge up the roof and the other edge resting just over the top edge of the first bundle. This "locks" the two bundles in place. Two bundles are laid in the same fashion on the opposite side of the caps. Three bundles are then laid across the caps and rest on the upper surface of the two "locking" bundles. The first bundles and locking bundles should be spaced so that the first layer of three bundles bridges across the cap and lies flat on the locking bundles. The second layer of three bundles should then be laid lengthwise along the caps over the first layer. Alternating the direction of the layers of shingles "locks" the pile. Four three-bundle layers over four locking and leveling bundles is plenty of weight in one place.

[(4 x 3) + 4] bundles x 75 lbs./bundle = 1,200 lbs.

It would take an extremely high wind to displace bundles stacked in this manner.

Nail: An uncaring supplier or contractor can nail you by stacking your shingles wrong. The shingles can be stacked by just piling all the bundles across the cap and letting the bundles bow down into a U-shape. It is tougher to lay these bowed shingles and get a nice finished job.

There is an extra fee for loading the roof as opposed to stocking everything on the ground. In our area the suppliers charge $1.00 to $1.50 additional *per square*. If your house is Figure 1, you will need 22 squares or 66 bundles on your roof.

66 bundles x 75 lbs. per bundle = 4,950 lbs.
4,950 lbs/2,000 lbs./ton = 2.48 tons

It's worth the $22.00 to save lugging two and a half tons, 75 pounds at a time, up a one- or two-story ladder. If the supplier doesn't have a way to deliver to your roof, find another supplier.

4.
Work to Do Prior to Roofing

PAINT

You may have work to do before you lay the first shingle. Do the dormers or divider walls above sections of your roof need to be painted? Paint first and save yourself the risk of splattering your new shingles. Also, paint any metal chimneys or metal vents prior to roofing. (Auto supply stores have aluminum and black "high heat" paint in spray cans. It stands up well on the outer metal surface of double-wall metal chimneys.) Painting before you lay the new roof means you will overlay or tear off any overspray or drips.

REMOVE OVERHANGING LIMBS AND TREES

Remove limbs that can damage your roof during high winds or snows. Remove any leaning, dangerous trees before you reroof the house.

CUPOLAS

Cupolas are small, usually decorative, venting structures. Remove, repair, and paint the cupolas. If the cupola is functional and removing it leaves a hole in the roof, patch the hole with plywood and roofing felt or polyethylene. Replace the cupola when the new roof is complete.

CLEAN GUTTERS AND DOWNSPOUTS

Cleaning gutters and downspouts can be a messy job. Go ahead and give them a basic cleaning before you begin laying shingles. An overlay will narrow the space between the lower edge of the new shingles and the outer lip of the gutter. This will make a gutter full of leaves and dirt harder to clean. If you are going to tear your roof off, it is still a good idea to clean the gutters first. You want the gutters to catch as much stuff as possible as it rolls off the roof. You will be amazed at the nails, grit, and pieces of roofing that the cleaned gutters will catch. You're going to clean the gutters when the job is completed anyway, and it's much easier to clean the gutters a second time than to pick nails and other trash out of the flower borders and yard.

REPAIR CHIMNEYS

If you have a brick chimney, check its top. Inspect the concrete chimney cap around your flues carefully. If the cap is cracked, use silicone caulk or roofing mastic to seal the cracks. If the cap is shattered or loose, chip it off with a hammer and chisel. A premixed concrete such as Sakrete is good for a chimney cap. Add water and mix it a part of bag at a time in a wheelbarrow. Go easy on the water; make the concrete thick or stiff, not soupy. Carry the concrete up the ladder a couple of gallons at a time. (A 5-gallon bucket full of concrete is too heavy to fool with on a ladder.) Trowel the new cap in place. Taper it down from the flue liners to the outer brick faces of the chimney. Keep the concrete moist as it cures over the next twenty-four hours.

The chimney cap was thin and badly cracked.

New cap in place prior to new roof. Do the cap repair on a cool, cloudy day and keep it wet as it cures. (Don't spray directly with the hose or concrete will fly everywhere. Lightly splash the water on by hand.)

5.
Overlaying the Original Roof

If your home has the original (or only one) shingle roof on it, the building code allows you to overlay it once with a new roof. If your roof has already been overlaid once, you must tear off both the overlay and the original roof down to the plywood or plank sheathing.

Weight is a key reason for the maximum limit of two roof systems on a home. What does the roof in Figure 1 weigh?

22 squares = 66 bundles: 66 bundles x 75 lbs./ bundle =
4,960 lbs.: 4,960 lbs./2,000 lbs./ton = 2.48 tons

Overlaying this roof will double the amount of the permanent weight, or "dead load," to a total of five tons. This is within the code. If someone disregards the code and puts on three roofs, they now have seven and a half tons overhead. Ice or snow will add tons of weight ("dead load"). Wind causes loading, but the wind is shifting and variable so it is considered a "live load." Seven and a half tons of roofing material and a few tons of snow and ice combined with the "live load" from a blast of high wind can cause a roof to fail without warning.

The other key reason for only two roofs is that you want your roofing nails to penetrate the sheathing to give you maximum hold. If you try to put on a third

roof, the shaft of your roofing nail has to go through three overlapping layers of shingles and may just barely penetrate the wood sheathing. When you nail new shingles to nothing but old shingles, a high wind will peel your new roof like a banana.

Having said all that, overlaying the original or a single roof is by far the preferable situation in terms of labor, mess, and risks due to weather.

Nail: An unscrupulous contractor may try to talk you into an overlay even if you already have two roofs on your home. If you are dealing with someone who isn't licensed and isn't insured anyway, what does he care?

Nail: A favorite ploy of unscrupulous contractors is to "cut the tabs" on a shingle roof that is badly curled. This is cutting and removing the exposed portion of each course of shingles. They simply go along with a knife and use the bottom of the exposed tabs of each row of shingles to cut off the exposed tabs of the course down the roof. They cut their way up the roof, and all they have to do is throw away a bunch of 5" x 12" shingle tabs with no nails in them. They don't have to replace the felt, pop new chalk lines, treat the valleys, etc., so they can naturally discount the price to the consumer ever so slightly.

Some homeowners spring for this kind of deal, but remember, if the tabs are cut on an existing second roof, less than ¼ of the overhead weight is removed, and there is still the problem of full nail penetration through the sheathing.

DON'T PUT IT OFF

Procrastination does not pay when you are deciding to overlay your original roof. If your original roof was laid using twenty-year shingles, start keeping a close eye on it at the fifteen-year mark. While it is true the roof can last twenty years, the shingles at twenty years may be so brittle, curled, and broken up that you can't overlay it.

A brittle roof will crunch like corn flakes when you walk on it. Don't try to overlay a roof that is brittle. Small chunks of shingle and grit will roll down the roof as you work on it, and they will get between and under your new shingles unless you constantly brush away the pieces that roll down the roof. It will take less time and be less aggravating to go ahead and tear off an old roof like this.

If you look up the pitch of your roof from the ground, and your shingles look swollen in sections or have a "bear claw curl," the unevenness will not only show through your new shingles, but you will still contend with the broken pieces rolling under your new shingles. My advice again is to tear it off.

Nail: Contractors who don't own a dump truck are especially prone to push for an overlay, no matter what the condition of your original roof.

When a roof is torn off and the shingles and asphalt felt underlayment, or "paper," are thrown in the back of a pickup and sit in the sun for several days, everything tends to weld together into a single nail-and metal-laden mass. Think about trying to get that stuff off by hand. When we emptied our big dump truck, we usually had the tear-off from three houses in it. Despite being dumped and occasionally dragged, the material stayed in the perfect shape of the dump bed. If you are doing a tear-off, don't let the trash sit long enough to bond.

BASIC INFORMATION ABOUT SHINGLES

The standard modern shingle is a rolled laminate of fiberglass matting and asphalt with a surface of various colored grit. The top of the shingle is solid across its entire 36-inch width. The grit on the top is basically black. The bottom of the shingle will be the exposed portion, and the grit on the bottom gives the shingle its color. The bottom of the shingle is notched out at 12 inches and 24 inches with *keys* that are ⅜ inch across and 5⅛ inches long. Each end of the shingle is trimmed to form a half-key. The keys form three "tabs" in the bottom of the shingle. You will make your measurements and lay your courses based on the top, full 36-inch width, of the shingles.

The tabs allow the shingles to be flexible, so they will lie flat over imperfections in the original shingle roof or sheathing underneath. The tabs also limit the amount of damage from severe winds. The shingle is cut into tabs so that a high wind can tear the tab off. It is better to lose a fairly small 12-inch tab than a full 36-inch shingle.

The manufacturer cuts the shingles within certain fairly small tolerances (usually +/-1/16 inch). However, trouble with odd-sized shingles does occasionally happen. The first time I saw this, the *keyways* (vertical alignment of the keys up the roof) on the house I was roofing were waving all over the place. The last thing I checked was the width of the shingles themselves. I found some to be not 36 inches but 36½ inches across the top.

Nail: A poor contractor is not going to worry about whether your keyways are straight. A shoddy contractor's keyways aren't very straight even if his shingles are cut perfectly at the factory. He will try to convince you that straight keyways aren't important.

NAILING SHINGLES

Nails should be driven above the keys and just below the asphalt self-sealing strip. Nailing in this spot ensures that the tab for the next course up the roof will center over the nailhead and cover it fully.

Also the pressure from the nail tends to raise the self-sealing strip, which makes it seal more fully to the shingle above.

If you are nailing into plank sheathing, you may find that you nailed directly over a gap between the planks. If you already drove a nail that didn't "bite," leave that nail in place so there won't be a hole. Nail the shingle again, keeping your nails up higher above the gap in the planks. Now nail your new course higher. If necessary, stay above the self-sealing strip with your nails. Don't go any higher above your optimum nailing position than you have to, but make sure you nail into solid wood.

Your nailhead should be driven down snug with the surface of the shingles. Don't drive the head into or through the shingle. (If you use a nail gun, be especially careful to set your air pressure low enough so the nailheads don't tear into the shingle.) When the head goes through, the asphalt has been cracked and the fiberglass torn. It's a weak spot. It may never leak, but why take the chance? If you are using a gun, carry a claw hammer with you to drive down the occasional nail the gun doesn't drive flush.

METHOD

Now get roofing! The first thing you need to do is cut off your shingles where they protrude over the edge of your roof or *rake*. The trim board that runs up the side of your roof from the gutter to the caps is called your *rake board*. Cutting off the protective overhang of your old roof is called *cutting the rakes*.

On most homes, the carpentry on the rakes is straight, but as a precaution, sight up your rakes from the ground. Are the rake boards straight? Are they bowed in or bowed out? If they are bowed in toward the house, it's not a problem as far as re-roofing is concerned. If the rake is bowed in more than half an inch, you need to decide if you are going to cut the rake off the old roof, following the bow in the rake board, or if you will follow a straight line up the roof to support your new shingles over the bowed area.

Go to the rake, just above the fascia. Using your hook blade, trim the shingles back gradually to the face of the rake board. Now make a short slice up the roof and all the way through the shingles, making sure it is even with the face of the rake. Go to the peak of the roof just below the cap pieces and again gradually notch back to the face of the rake. Go back down the roof and hook the metal nail hook from your chalk line securely in the short slice you made. (If you don't have a metal nail hook, tie the end of your line to a standard metal washer.) Now reel your chalk line out. (You may need to bang the side of the box on your hand to loosen the chalk powder as the line comes out.) Pull the chalk line taut and hold it directly above the face of the rake board so that you can see through your notch at the caps. Holding the line firmly with one hand, raise the line straight up and "pop it" (release it). You have just "popped a line" above the face of your rake board.

Now you need to cut the rake. It is easiest to start at the top and cut down the rake. Tuck one leg and use your other leg as a brake and brace. Use the hand closest to the roof to hold the top outside edge of the shingle to be cut. With your other hand, slowly pull the hook blade down the blue chalk line on the top shingle. Cut the upper portion of the next shingle down the roof using the edge of the shingle you just cut as a guide for your blade. Finish cutting this shingle, using the chalk line to guide your blade. *Don't cut toward your hand or your body — ever!* Lay the pieces you trim on the roof above you and collect them when you finish cutting the rake. It's easier to pick the pieces up off the roof than to fish them out of a flower border or shrubs. When you get close to the bottom of the rake, turn around and pull your hook blade up the roof.

If you have tough knees, you can just squat with your back up the roof to make this cut. If you are like me and want to be able to walk the next day, tuck one leg under and plant your other foot firmly on the roof, using your leg as a brace and the sole of your shoe as a brake. If you are right-handed, tuck your right leg and get it out of your way. Be careful not to pull against the knife so hard that you pull yourself off balance.

Nail: Too many contractors "cut the rake" on the old roof with the hatchet end of a roofing hammer. They hack the overhang off along a jagged line an inch or so behind the face of the rake board. New shingles won't stay straight over this hacked area plus the required 1-inch overhang beyond the rake. The edges of the new roof soon curl in a pronounced downward hump, which begins before the shingles even reach the rake board.

Your new shingles will be strong and come straight off the lower edge of your roof (fascia) above your gutter. Your old shingles may have drooped down and become ragged where they overhang the trim board covering the end of your joists (fascia board). To make your new roof look neat, you need to cut along the fascia. Repeat the same notching process at both ends of your fascia and pop a line across the lower edge of your roof. If your roof is extremely long, you may have to cut another notch in the middle and pop the line in two steps.

Now you need to "cut the third course." Let me give you some background on this. Your roof is already two shingles thick at its lower edge. Your new roof will also be two shingles thick at its lower edge. This means your new roof will be four shingles thick at the fascia. This extra thickness will create an unsightly hump where the extra thickness changes down to a normal thickness. To eliminate this hump, you need to taper the thickness of shingles gradually from four back to two. You do this by cutting the third course.

Your shingles are 12 inches high, and you will let them overhang your gutter $1\frac{1}{2}$ inches. The top edge of your new shingles will be $10\frac{1}{2}$ inches above the fascia cut you just completed. (12.0 - 1.5 = 10.5 inches)

Some roofers cut off the third course and use the cut edge as a guide for laying the new shingles. You should give yourself a little room and measure 11 inches up from the face of the fascia board and notch the shingles at the rake. Hook a chalk line in the notch and measure up 11 inches on the other end of the roof. Pop a line at 11 inches. Now cut the tabs of

the third course at the chalk line and remove the cut tab sections from the roof.

Nail: A bad contractor won't do any of the preparation and cutting covered so far. He will leave the ragged rake and fascia edges of the old roof showing under your new one. By not cutting off your third course, he will leave a pronounced and unsightly hump in your shingles. Some contractors who do cut off the third course, do exactly that. They cut off the third course even with the bottom of the shingles of the fourth course. This removes the entire 5 inches of the tabs of the third course instead of just the few inches at the bottom of the tabs that do need to come off. Removing the entire course causes a $2\frac{1}{2}$- to 3-inch gap underneath your new shingles. Your new shingles will try to bridge this gap, but will eventually dip down into it. Worse yet, your contractor will nail your new shingles down into this gap.

BOTTOM COURSE LINE

Allowing an overhang greater than $1\frac{1}{2}$ inches will cause a standard strength twenty-year shingle to droop over the fascia. Measure up your roof $10\frac{1}{2}$ inches from your fascia. Use your straight blade to make a horizontal arrowhead mark at $10\frac{1}{2}$ inches. Pull $10\frac{1}{2}$ inches at the other end of the fascia and drive a nail partway down at this measurement. Attach the nail hook on your chalk line and pop the $10\frac{1}{2}$-inch line across the roof.

BASE LINE AND OFFSET LINE

I don't recommend that you lay out your roof this way, but this is the way most roofers lay out the *verticals*, or vertical keyways up the roof. Your shingle is 36 inches wide, so if you measured in 36 inches from the face of your rake board and popped a line parallel to the face of your rake, the outside edge of your shingles will be exactly even with the face of your rake. But you have to have an overhang to protect the rake. A 1-inch overhang is generally accepted, but you will need to have enough shingle hanging out there that you can cut it easily with a hook blade. Allow yourself an extra inch to cut. Subtract the amount of the overhang from the 36 inches to get the measurement from the rake to the *base line.*

36" - 2" = 34"

At the top of the roof, just below the caps, measure in exactly 34 inches from the face of your rake. Mark the 34-inch measurement, using your straight blade to cut the point of an arrow in your old shingles. While your tape is still in place, mark a point 6 inches back toward the rake from your 34-inch base line measurement. This is the measurement for the *offset line.*

34" - 6" = 28"

Drive a nail solidly into the exact point of each arrow mark at the top of the roof, leaving part of the shaft of the nail exposed. Tie or hook a chalk line to the nail at the 34-inch mark. Mark the 34-inch measurement at the bottom of the roof. Hold the line tight over the 34-inch mark at the bottom of the roof and pop a line. Repeat the process for the 28-inch marks. You now have your base line and offset line to lay your shingles.

So far, this seems simple enough. Now I'm going to make things tougher for you. I don't want you doing it this way! The problem with just popping a base line and an offset line along the rake of the roof is that your error accumulates as you lay shingles across the roof. Your keyways as you progress toward your far rake become increasingly wavy. You may butt the ends of the shingles tighter one time and slightly looser the next. This causes a slight distortion of your keyway. Compounding the problem is the fact that shingles are not cut to microscopic perfection. They are the same size within certain tolerances (+/-$\frac{1}{16}$ inch). The acceptable variation in widths of the individual shingles will also cause your keyways to wave as you sight up your finished roof. If you pull the measurements for the base line and offset line close to the middle of the roof section, you cut your natural error at each end by half of what it would have been on the far end, if you set the base and offset at 34 inches and 28 inches and simply ran your courses the entire length of the roof.

Bear with me, because it's not complicated to get your lines in the center of the roof. Say you have a roof that is 35'8" from rake to rake. You want a 2-inch overhang at each rake so:

35'8" + 2" + 2" = 36'

Each shingle is 3 feet wide at the top.

36' length of roof/3' per shingle = 12 shingles

So about 12/2 = 6 shingles will put you at the center.

6 shingles x 3'/shingle = 18'

Remember the 2-inch overhang over your rake so:

18' - 2" = 17'10"

Go just below the cap pieces and measure in from the face of your rake 17'10". Mark it with your arrow using a straight blade knife. Now drop back 6 more inches to 17'10" - 6" = 17'4" and mark another arrow for the offset line. Do the same thing just above the fascia. Drive nails and pop the base and offset lines. By taking this little bit of extra trouble, you keep the error from compounding all the way across the roof and the keyways will stay straighter.

Figure 4 shows the layout of the bottom course line, base line, and offset line for this roof.

Most subdivision houses were designed with multiples of the 3-foot width of shingles firmly in mind. If you are in an older subdivision, older standard widths were in effect. If you are in a "one of a kind" home, your lengths of roof may bear no relation to a multiple of 3 feet. There will be examples of figuring odd sizes for you later.

Don't get overeager and start yet. Please read Chapter 17 about caps so you will understand why I recommend you use a subtraction amount that varies by $\frac{1}{2}$ inch to compute your base line and offset line measurements on the opposing side of the roof. After you have a little experience cutting shingles, you may need less than a full inch to cut off at the rakes. Maybe you can change the amounts you subtract to $1\frac{1}{2}$ inches and $7\frac{1}{2}$ inches on the opposing roof. The less you have to cut off, the stronger your tabs along the rakes will be.

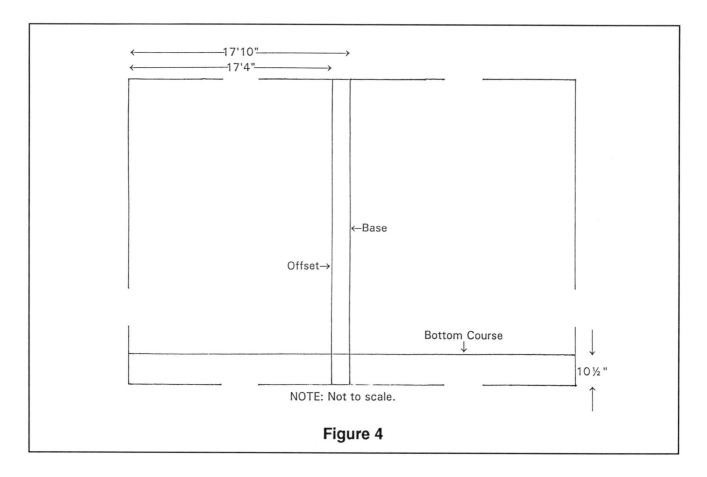

Figure 4

DOUBLE COURSE

Your "starter course" is the first course of shingles, which forms the unbroken lower edge of your roof along the fascia. The starter course shingles are actually turned 180 degrees, with the grit still facing up, the tabs lying up the roof, and the top of the shingle overhanging the fascia into the gutter. Line the side of a shingle exactly on the center of your base line, with the shingle itself lying in the direction of the rake from which you measured. Line the bottoms of the tabs on the center of the bottom course line (10½ inches above the fascia). Nail the first shingle in place. Continue toward the rake you measured from by laying your next starter course shingle beside the first.

Lay the first shingle for your "first course" over the starter shingles you just laid. The first course is laid in the normal manner with the keys pointing toward the fascia. *Center the end of this starter course shingle directly in line with the offset line, which is visible up the roof from the starter shingles.* The *starter* course provides an unbroken surface beneath the keys of the *first* course and also doubles the strength of the new shingles at the fascia.

This is a fine point, but you may want to consider it. It's a good idea to drop your first course down about ⅛ inch over the lower edge of the starter course. When your roof is completed, the water running off the shingles will tend to dribble back under the lower edge of your shingles. If the first course protrudes an additional ⅛ inch, the drips that try to dribble back will hit the edge of the underlying starter course and drop off rather than run back toward the fascia board.

The starter and first courses can be confusing when you first start roofing. Lay the starter course to both rakes, then come back and lay the first course.

Your final starter course shingle will overhang the rake you measure from by 2 inches. Your final first course shingle overhangs the rake with a long tab extending 8 inches beyond the rake.

LAYING THE ROOF

Some "roofers" make the mistake of jamming the shingles tight against each other, thinking it will make a "tighter" roof. Remember that shingles expand and contract and need a little breathing room just as people do. If shingles are jammed too tight against each other, they will tend to pucker or buckle.

One trick to keeping shingles from drooping over your rakes is to rough-cut the longer overhanging tab as you work up the roof. If you've got 8 inches sticking out over the rake, cut off 5 inches. That will keep the weight of the 8 inches from bending the tab down and possibly cracking the shingle at the rake. Once a shingle cracks at the face of the rake, it's going to droop from then on.

Nail: Too many contractors don't trim the rakes until the end of the job. You can see the overhanging shingles flapping in the wind, and they end up broken and hanging straight down. The contractor can come back and cut these rakes fairly straight, bending them back up with his hand, and they will look pretty good for a short time (at least until he can cash your check). Then the shingles are going to droop over the rake. The only way to correct the situation is to replace all of the shingles along the rakes. If you are lucky enough to get this kind of contractor back, he will try to convince you that the shingles drooping over your rake give your rake board added protection. You might as well believe him; he's not going to redo it.

One of the great advantages of an overlay is that you don't have to mark horizontal courses. Instead, you slide the top of the new shingle against the bottom of the tabs of the next course up the old roof. In effect, you are using the previous contractor's courses (hoping they are straight). Butting the top of your shingles up against the bottom edges of the course above is called "nesting" the shingles (as in bird's nest).

Now lay the second course up the roof. Line up the edge of the shingle with the base line and "nest" the top of the shingle against the bottom edge of the tabs of the fourth course of the old roof. Nail the shingle in place, being careful to keep your nails above the 1/2-inch gap between the top of your starter and first courses and the cut edge of the old third course.

When you come to the rakes, double nail the ends of the shingles. Nail once just below the self-sealing strip and nail again above that same nail, near the top of the shingle. Double nailing the rakes gives the roof tremendous extra strength in a high wind. Be careful that these end nails don't split through the face of the rake board. Also, make sure they won't show in the keyway of the shingle in the next course up the roof.

Lay the third course shingle by centering the end of it over the offset line and nesting the top of it against the lower edge of the fifth course of the old roof.

Lay the fourth course by lining up on the base line again and nailing across.

One costly and time-consuming mistake is failing to alternate courses between base and offset lines. For example, if two consecutive courses are laid up the roof, both starting at the base line, the roof is open at the sides of the shingles and lets water through to the old original roof underneath. That original roof is now peppered with nails from the new overlay and is not waterproof. Keep checking yourself as you work your way up the roof. You don't want to finish the section of roof and find that you failed to alternate courses down on the seventh or eighth course.

If you made this mistake, there is a way to fix it, especially if it's a roof that can't been seen. Carefully pull the nails on the upper of the two courses you failed to alternate. Slide a course of shingles between the two courses you messed up. If your two messed up courses both started at the base line, slide the correcting course in and start it 6 inches over at the offset line. Chalk line across your roof 2 1/2 inches up from the lower edge of the bottom course where you made the error. Nail your correcting course in place with its lower edge on the 2½-inch chalk line. Then nail your upper course back down.

If you discover this error on a roof that shows, and you don't want to roof the "mistake" again, get an answer ready. It will be a rare person who will be critical of the two courses (lower and correcting) that only have 2½-inch tabs showings. But if someone remarks on it, try responding with, "The roof looked so rigid and uniform. I wanted to customize it. Like it?"

I saw where one roofer discovered his mistake partway up the main roof of a home and then intentionally carried the same "short course theme" to the roof on a connecting wing. He then came down the same number of courses from the top and ran another short course on the main roof and wing. The owner was delighted with this very "distinctive" roof. Sometimes it's not what it is, but what you call it. If you're convincing enough, it may never be called a mistake.

The starter course and working position cause the first four courses to go slowly. Once the fourth course is laid, you can turn around and face up the roof. (If you aren't comfortable that close to the edge, go ahead and lay the fifth and sixth courses before you turn around. You aren't trying to prove anything to anybody.)

Most modern shingles are designed to give a completely random pattern. The manufacturers let the color of the surface grit vary across a 36-inch shingle. One shingle may have one end darker than the rest, and the next shingle may have the center darker than the rest. The random pattern serves to hide natural wear, scuff marks, and staining as the shingles age. Patterning also relieves the manufacturer of trying to produce a perfectly uniform color.

Patterned shingles are designed to be laid across the roof, but many roofers lay the shingles straight up. It is quicker and easier for them to get on their knees and the bottom of their toes and lay straight up the roof, starting at their base and offset lines. As I said earlier, most roofers have the base and offset lines 34 inches and 28 inches from the rake. They go straight up the side of the roof, using the base and offset lines. They carry the same bundle of shingles right up the roof, pulling shingles from the top of the bundle as they go up.

Laying shingles straight up leads to "tiger-striping" of the shingles. When the roof is finished, you will see a broken striping effect as you look at the roof from a slight distance. Darker lines angle up the roof.

Laying straight up is also obvious when the roof is a color without a random pattern. Certainteed makes a white shingle that is pure white. Remember what I said about different production runs being different shades. A home in our area faces away from the main road. The back roof shines a shimmering white except for one cream-white strip rising vertically near the center of the roof. The roof looked as if it were patched the day it was done. It's surprising how many people have noticed it.

My method of laying shingles *across* the roof breaks up the "tiger-striping." I stock roofs, breaking the bundles of shingles up as I go. For example, if the section of roof you are going to overlay has an area of ten squares, you know you will need to stock thirty bundles (10 sq. x 3 bundles/sq. = 30 bundles). Space these bundles evenly on the roof. Tear the paper off the lower bundles on the roof, but leave the upper bundles wrapped until you are ready for them. (A wrapped bundle will withstand unexpected high winds.) Split the bundles in half as you need them and skip the halves across the roof. It takes a few extra minutes, but your finished roof will have a pattern that is absolutely random.

Nail: An unscrupulous roofer will try to convince the customer that the manufacturers prescribe laying the shingles straight up the roof. As you know, this is not true. You know about "tiger striping." Another problem is that roofers laying the shingles vertically don't usually butt one vertical course snugly against the previous vertical course. This leaves extra wide keyways where the vertical courses meet. These extra wide keyways are obvious from the ground.

A more serious problem occurs when the roofer jams the butt ends of the shingles together on vertical courses. When the shingles are jammed, there is no room for the shingles to expand and contract. On a hot day jammed shingles tend to pucker up into "fish mouths."

FINISHING DETAILS

When you finish a section of roof, there is one thing you should do immediately before the sun seals the tabs down. Have someone stand on the ground and sight directly up the roof. Have them direct you to every "fish mouth" and gently lift the raised tab. You will find either a raised nail or a little piece of trash holding the shingle up. Either snug the nail down or brush the scrap of trash out of the way so the shingle will seat and seal down. A fish mouth is a weak spot because it gives a high wind a place to start tearing off your new roof.

You can help the person on the ground by standing to the side of the immediate area he is checking and holding the head of your hammer down on the roof as he directs you. The vertical handle of the hammer gives him a reference and helps him tell you which direction to go to find the flaw.

Nail: Some roofers who still "hand nail" their roofs don't make sure their nails are down snug. However, the problem of "fish mouths" is more prevalent with an air gun. Too many roofers don't take the time and trouble to snug all the nails (staples) down and clean the trash from under the tabs.

CUT RAKE OVERHANGS

If you set the roof up as described earlier in this chapter, you are going to have one course of shingles overhanging the rake by 2 inches. The longer tabs will overhang the rake by 8 inches. The long tabs are heavy: if you leave them there they will sag over the face of the rake. Sagging weakens the shingles, and a high wind will make the overhanging tabs flap and break along the edge of the fascia board. You need to cut some weight off the long tabs. Use your hook

knife and cut off 4 or 5 inches. I find it best to cut off the long tabs as I lay the courses up the roof. Giving the long tabs this rough trim removes extra weight and allows your shingles to continue sticking straight out from the rake.

Trim the rakes as you finish each section of roof. Trimming as you go will keep the overhang from sagging, and you won't have to worry about trimming the whole roof in the event of a sudden storm.

You need to pop a chalk line to trim the rakes. Start by sitting at the bottom of the rake board. Stick a 25-foot metal tape under the overhanging shingles and butt the end of the tape against the rake board. Begin trimming the shingles back until the metal tape shows you have exactly 1 inch of shingle overhanging the rake. Make a single cut up the roof at this 1-inch mark. Hook the metal nail clip at the end of the chalk line beneath this cut, and lay out your chalk line on the roof. Cut back to the 1-inch mark at the top of the roof then gently lay your chalk line in position. Pull the line tight and pop a chalk line.

As when you cut the rakes off the old roof, each shingle will cover the upper portion of the next course down the roof. There will be no chalk on the top 7 inches of the covered shingle. Simply angle your hook blade back in toward the rake and use the cut edge of the shingle above as your guide. You should angle the blade because the cut you make using the chalk line is the straightest: it's the cut you want to be the most visible from the ground. When you look up your rake from the ground, it will look as if you cut it with a laser.

Nail: Trimming the rakes is one distinctive feature that is easily seen from the ground. There are all kinds of ways to trim a rake. Some roofers won't take the time to pop a chalk line for their rake. They just use the first joint of their forefinger to measure the overhang as they trim down the roof. The rake edges on roofs done this way look like an alligator chewed them off.

Lightly sweep the roof down. The asphalt in the small scraps tends to melt and make the trash adhere to your new shingles. Don't try to sweep every bit of loose grit. You will end up breaking loose some of the grit that was still bonded tight. Just get the small scraps and some of the loose grit off the roof.

Sweeping each section (into the gutter) as you go gives the roof a finished appearance and makes the final cleanup much easier.

When do you replace a roof? The shingles on the right have cracked, and a nail in the lower left has popped through. The shingles crumble slightly underfoot. It's time.

These are the shingles on the main roof of the eager do-it-yourselfer before he overlaid it. It's past time.

Corners are curling.

Severe "bear-claw curl" plus damage from foot traffic as crew loaded new shingles.

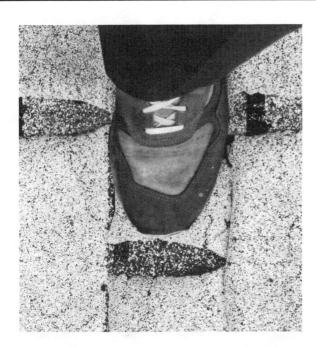

Damage can be minimized by walking with the shingle courses and keeping your weight on the edge of your shoe at the top of the shingle.

Nail: Any roofer who looks at your roof wants the work. A shoddy contractor will club-foot and drag his shoes. You will definitely need a new roof when he comes down the ladder.

Nail: The nail here is that the contractor didn't double-nail (or apparently single-nail) the rakes. The odd color patches are easy to spot. Look closely to see the large patched area between a quarter and halfway up the rake.

Here is a "distinctive" roof on a $275,000 home. Look eight courses up.

Shingles laid straight up and not snugly butted.

"Fish mouths" occur when partially driven staples and pieces of trash keep the shingles from lying flat and sealing. These "fish mouths" are on a model home in an upscale development.

Rake on the same model home. The rake cut is rough, plus the keyway is extremely close to the rake. The "pigtail tabs" will soon be drooping over the rake.

6.
Tear-Off and Structural Repair

The methods for tying in and flashing various roof penetrations and structures are similar for an overlay or a tear-off. The difference is that the overlay often requires only the final few steps of the several that would have been required if you were working on a complete tear-off (or new roof). If you are overlaying your roof, it is helpful for you to know what is underneath your old shingles. The knowledge of how the work was done originally will help you understand why you are doing certain things on your overlay. In the remaining chapters, you will see how to complete a tear-off roof.

Let's review the reasons an old roof should be torn off:

1. There are already two complete roofs on the house, the original plus one overlay. If your home is in the forty-year age range, chances are good it already has two roofs.

2. You only have one roof, but the shingles are badly curled and crumbling.

3. Inspection from inside your attic reveals extensive deterioration of your sheathing and/or rafters.

4. There are multiple problems with the roof, including poor workmanship and waving, erratic alignment of the shingle courses.

5. You only have one roof but it is done in dimensional shingles. Dimensional shingles simulate the coarse texture of cedar shakes and don't have an even surface for you to overlay.

There is too much work, risk, and mess for even the most extreme perfectionist among us to do a tear-off because, "It's the way to really do the job right." If your roof doesn't need to be torn off, a well-done overlay is great.

My number one rule for a tear-off is: never begin a tear-off until all your materials are stocked on the job. There is a story behind this rule. My main supplier was absolutely and completely reliable. My materials were always on the job on time or early, except once. That one time he promised me absolutely that the shingles would be there by midmorning. I started tearing off the rear roof of a customer's home. There was still no sign of my supplier at 10:00 AM so I called him. One of his drivers had called in sick, but everything was okay. By late morning the rear half of my customer's roof was gone, and black clouds were rolling in despite the forecast of perfect weather. I called my supplier again. One of his trucks had broken down and his third truck was stuck in the mud on a new construction site. That stuck truck was the one that was supposed to bring my supplies to me. My materials were still sitting in his warehouse. I dumped the bed of my truck in the customer's driveway and dashed through heavy lunch hour traffic across two counties and picked up my "delivery." My crew had been sitting idle for two hours when I got back to the job. We got the No. 15 felt down over the plank sheathing just as the rain hit. Fortunately, it wasn't a blowing rain.

My risking a tear-off before all supplies were on the roof almost led to a disaster, but you can learn from my mistakes. I don't care what promises, commitments, time constraints, or whatever else you have, don't tear off until everything you need is sitting there ready to go.

Nail: Some roofers habitually start a tear-off with no materials on the job and no intention of getting the house watertight that same day. I have seen houses with roofs covered by tarps over the weekend, and I've passed houses with half the roof gone in the middle of a rainstorm. Some roofers have the attitude, "That's why I have insurance." Others have the attitude, "I'm not insured anyway, so if it gets wet inside, I'm gone." Either attitude leaves you with a damaged home and furnishings. Don't allow your roofer to tear off if his supplies haven't been delivered.

If your friends are willing and able to give you a hand with the tear-off, don't be too proud to accept help. It's hard work and can leave you worn out before you ever start reroofing.

Since you are new at this, pick the smaller and easier section of roof for your first tear-off. If you are reroofing your house and detached garage, start with the garage. Ease yourself into this tear-off so you develop a feel for it.

If you have a section of roof with no valleys, skylights, chimneys, etc. and it also is the farthest roof from your dumpster, truck, or trash pile, start there. By going to the most remote site and working toward your trash pile or truck, you won't have foot traffic over a new roof and run the risk of scarring it.

You can tear a roof off using a square shovel if you want to. It gets under the shingles and will pry them up and pull most of the roofing nails out too. Better tools are available. A shingle stripper (roofing spade) is a narrow square shovel with large V-notch teeth cut in the front of the blade. The teeth will go under the head and around the shaft of a stubborn nail and make it easier for you to pop it out. A pipe is welded across the upper back of the blade so that when you push the handle down, the pipe acts as a fulcrum and gives you leverage to pop the nail out. A short-handled shingle stripper is best on a steep roof (greater than 8/12). It lets you work right beside where you are leaning against the roof. A long handled shingle stripper is best on a moderately steep roof (6/12-8/12); it lets you reach farther.

There is a tool called a *shingle eater* that has a wider blade, a sharp upward angle at the rear of the blade to act as a fulcrum, and a metal handle that is bent upward. The sharp upward bend of the handle saves your back by letting you stand more erect as you tear off a 4/12 or 5/12 roof.

It depends on your preference whether you want to buy these specialty tools. You can do the same job with a square pointed shovel and a claw hammer, it just takes a lot more effort.

METHOD OF TEAR-OFF

If your sheathing is plank, cover anything valuable in your attic. Grit, nails, scraps, and dirt are going to fall through the gaps in the plank.

Start your tear-off across the ridge from the roof you want to remove. Insert your stripper beneath the bottom edges of the shingles two courses down. Pry these shingles up and keep working your stripper up toward the ridge of the roof. Come on across the ridge, prying the caps up and free. Throw all the old caps and shingles in your truck or trash pile. Now you are ready to start down the side you are going to tear off. Get your stripper under the *felt* and start prying the felt up along with the shingles as you work your way back and forth across and down the roof.

Some roofers like to tear off coming up from the bottom of the roof. Tearing off going down the roof gives you bigger chunks of roofing that don't separate into individual shingles. Going down the roof helps keep all the scraps, grit, and nails down below you while you keep good footing on fairly clean sheathing.

Nail: Some roofers try to talk the customer into reusing the felt from the original roof. They will very carefully tear off the shingles only. Reusing the old felt saves the cost of the new felt. Since the original roofer's old lines will still be visible, it also saves the time it takes to mark the courses and verticals (base and offset lines). The old felt won't act as a backup roof. It will turn away about as much water as Swiss cheese.

Your supplier may have stockpiled shingles on the ridge. Don't try to move the piles; just tear off around them. You want to move the bundles in this pile around as little as possible. You want to move the bundles from the pile to where you are going to use them. You don't want to double your work by moving the whole pile to another location on the roof and then later have to bring the bundles back to where you are going to use them.

Keep your nails and grit swept off the exposed sheathing as you go. The trash can be as slick as ball bearings. When you get to the bottom of the roof, let the trash fall into the gutters as you roll the tear-off material into large chunks.

Be especially careful to keep a clean footing along the bottom of the roof. (A wide shop-type broom works well on a roof.) I only had two men fall. Both times they were tearing off. Both times they were at the bottom of the roof. Both times the roofs were 5/12 (safe). Both times the men had let the trash accumulate under their feet. Neither man thought the grit and nails would throw them. Neither man had time to yell before he went over the edge. Both men were lucky. One fell in a dense thorny bush which broke his fall, but he looked like he had been attacked by fifty furious cats. The other man landed on a concrete walk and only lost a little time from work. He was bruised and stiff for days. Both men were very careful about keeping the roof clean after they fell.

SHEATHING REPAIR

You will snag and pull some of your sheathing nails as you tear off. Once you have swept the roof clean,

drive down or replace the sheathing nails. You also may need to add nails that the carpenter missed on the sheathing when your house was built.

Nail: Some roofers don't replace the sheathing nails they pull out or go back over any of the original carpenter's "skips."

If there is rotted or broken sheathing, replace it. You will find that when a sheet of plywood has a rotten spot, it is easier to tear the whole sheet out and replace it with an entire new sheet. Cutting out and removing a part of a sheet and then fitting a new piece back in is tedious and time consuming. If the rotten sheet had cutouts for vent pipes and such, use your old sheet as a pattern for the cutouts on your new sheet.

If there are major breaks in the plank sheathing, replace the plank as necessary. Plank sheathing may have broken due to a large knothole, or one of the planks may have cracked. If the hole or crack isn't huge, cover the hole with a metal patch. Cut a length off your aluminum coil and nail it over the hole. (An old piece of galvanized metal is good for this, too.) You are covering the hole to keep someone from stepping down into the hole and punching a hole through the new shingles in the future. The metal patch is an acceptable repair, and you can roof right over it.

Nail: On one tear-off, I discovered a 10-inch diameter hole in the plywood sheathing. The hole was either cut by mistake, or the plans were changed and the furnace, with its metal chimney, was relocated. Anyway, the original roofer laid the shingles right across the hole, with no metal patch. I was glad I hadn't stepped on that particular spot.

If your plywood sheathing has sagged between the joists, cut a 14½-inch or 22½-inch piece of 2 x 4 to fit the space between the joists. Pre-nail a couple of angle nails at each end. Push the plywood sheathing back up even, using your shoulder against the 2 x 4 cross brace, and nail the 2 x 4 cross brace into place. Once you have the cross brace angle nailed, you should end-nail the brace. Go to the sides of the rafters opposite the brace and nail through them into the end of the cross brace.

RAFTER REPAIR

If at all possible, get in the attic early in the morning before it gets too hot. If a rafter has sagged badly, loosen and raise the sheathing to push the roof surface back into shape. Push the sheathing up with a 2 x 4 that spans the length of the sag in your rafter. Face along the top of the rafter with the new 2 x 4, which you nail in place in its raised position across the length of the sag. (One person can do this, but it's far easier with two.) This will get rid of the dip that showed in the roof.

Face any broken rafters with 2 x 4's nailed on both sides.

If a rafter is somewhat dry-rotted and is questionable, face over the dry-rotted area with a 2 x 4 on one side of the rafter. Make sure your 2 x 4 spans over the rotted area and is nailed into good wood.

If a rafter is truly rotten, face one side of it with a 2 x 4. Remove the sheathing above and cut out the rotten section of rafter. Face the opposite side of the gap in the rafter with another 2 x 4 and restore your sheathing.

Once you have torn off, getting a roof back on your home is your number one priority. No. 15 felt does a pretty good job of waterproofing *plywood* sheathing. It doesn't waterproof *plank* sheathing quite as well. The problem with plank sheathing is that you are going to hit the gap between the planks with your swing tacker and tear a hole in the felt. Consequently, felt on plank turns *most* of the water. There is almost nothing you can do about it except push yourself to get the new shingles laid. There is *one* other thing you can do, but it hurts my pride even to say it. Polyethylene sheets spread over the insulation in the attic are a good backup system if you have just felted in the plank sheathing and have to ride out a storm.

Naj Altefmans the "shingle eater." Matt Porter carries off the tear-off material.

A piece of step flashing makes a great gutter scoop.

Naj Altef, Matt Porter, Doug Merold, and Darren Porter tear off my house. Note we left the shingles around the chimney and protected the aluminum awning with a tarp nailed to the bottom of the roof.

When you tear off the lower portion of a roof, it is preferable to tuck the top course of felt under the lower edge of old shingles of the section of roof you left at the top. If it isn't possible to tuck the paper under, seal it down well with mastic.

In the foreground, the old roof has been torn off and new dimensional shingles laid in place. The Number 15 felt overlaps the edge of the old shingle roof a few feet, and extra white shingles have been placed vertically over the felt to hold it down. The white shingles are nailed into place through the old shingle roof. On a 7/12 roof there is no such thing as placing a bundle on the felt to weight it down. The bundle will slide right off the roof. You put down one shingle and nail it in place through the felt to keep your felt from blowing loose.

The new roof is already in place in the front. Gently tear off the rear roof onto the front roof.

7.
Accessories

You can add trim pieces and additional features to your roof. Most of these would be added at this early stage of laying your new roof. Therefore, this is the logical spot to talk about roof "accessories."

Once you have torn your roof off, you will proceed in the same way you would if you were roofing a brand new home under construction. It's almost the same thing; there are only two differences. The first difference is the home below is finished and full of furniture and carpet. The second difference is it all belongs to you. Let me show you what the accessories are and then let's get busy laying your new roof.

GUTTERS AND DOWNSPOUTS

If replacement of gutters and downspouts were a part of my contract, I would replace the gutter at this stage. However, there's a big difference between working with a crew and working alone. Installing gutters isn't that difficult, but what if you discover rotted fascia boards that need to be replaced? If you are working alone, just give your old gutter a rough cleaning and replace it after your new roof is finished.

DRIP EDGE

Often when I installed new gutters, my customer also wanted an aluminum drip edge. This is an aluminum trim piece that is installed over the lower edge of the sheathing. The long leg of the drip edge lies up the roof on top of the sheathing. The short leg drops down and protects the bottom edge of the sheathing and the top of the fascia board. When a home has a gutter, the short leg of the drip edge drops down into the gutter. The short leg has to be notched out so it will drop down over the spikes and ferrules (long nails and spacer sleeves around the nails) that hold the gutter in place.

This requirement to notch the drip edge for the spikes and ferrules was the reason I liked to install the new gutters before I installed the drip edge. However, you can install the gutters, then notch the drip edge and slide it up under the No. 15 felt and new shingles, after your roof is laid. I know what you're thinking. Don't, *do not*, install new gutter before you tear off. There is a high probability that the tear-off operation will scar or damage the gutter.

Whether you are going to continue to use your old gutter or have gotten your new gutter in place, here is the way to do the drip edge. Lay the drip edge out along the lower edge of the roof and scratch marks in the short leg so you can notch for your spikes and ferrules. It is neater to make a vertical cut on each side of the spike and ferrule and just raise the flap of the drip edge up, leaving it in place resting over the spike and ferrule. You are less likely to kink the drip edge leaving the flaps uncut. (When you clean your new gutter several months from now, just don't drag your hand across the top of the spikes.)

The upper leg of the drip edge is going to be covered by the No. 15 felt, so you can use a standard galvanized roofing nail on this particular aluminum.

Don't overlap the ends of the sections of drip edge. It will cause a hump in the lower edge of your shingles and be visible from the ground. Just butt the ends together and nail them in place.

You should consider using drip edge if your roof has a fairly low pitch and water has been dribbling back in under the shingles and is starting to rot the sheathing or joists. Other than that, the drip edge is completely optional. Most homes don't have it.

ICE SHIELD

Certain parts of the country are subject to ice damming. This occurs when a wet snow partially thaws and refreezes, causing ice to form in the gutters. When the snow begins to melt over the main roof (where some heat is escaping from the attic), the water runs beneath the snow to the ice dam at the gutters. The ice dam backs the water up the roof. Shingle roofs are designed to handle water that is running downhill, not standing water. Standing water runs under the shingles and under the roofing felt into your home. There are some recent products on the market designed to counteract the problems caused by ice damming. I am going to call all these products by a name that is becoming generic — *ice shield*.

Ice shield comes in a roll usually 30 inches wide with a waxy paper backing protecting the sticky side. Ice shield goes directly down on your sheathing. If you use a drip edge, ice shield rolls out over the upper leg of the drip edge. You roll off the paper backing as you unroll the ice shield along the lower edge of the roof. Don't let any ripples get in the material because the stuff adheres instantly to the plywood. If you do get a wrinkle, just hammer it down smooth. Ice shield not only adheres but melds into the wood when your roof is heated by the sun. The top surface of some brands of ice shield is extremely slick, so for safety's sake, cover it over with felt. The ice shield will seal around the nails or staples.

Ice shield is not as effective on plank sheathing due to the wide gaps and butted end joints between the planks.

ROOF EDGE

There is another type of aluminum trim called *roof edge*. This runs up the rakes over the No. 15 felt. Roof edge is designed to handle windblown rain coming under the edge of shingles at the rakes. Rain blows in beneath the shingles and over the long leg of the roof edge. The moisture then runs down the roof over the No. 15 felt. As you lay roof edge up from the bottom of the roof, you should theoretically allow the next edge piece up to overlap the lower one by about 1 inch. (This keeps the joints lapped correctly for the water running down the long leg of the drip edge.) Where there had been a problem of water encroachment under the shingles, I did overlap the roof edge. However, on some homes the roof edge was not necessary in the first place, and the overlap was unsightly, so I just butted the ends together.

One small tip. Drip edge is often cheaper than roof edge, and some versions are made basically the same way. If you want both, try using the drip edge pieces to do both jobs.

Nail: Drip edge and roof edge are relatively inexpensive and easy to install. They end up being a "high markup item" for some contractors.

SEQUENCE OF INSTALLATION

It is crucial that you install the drip edge, ice shield, No. 15 felt, and roof edge in the proper sequence, or these items don't work and could even become a liability. Let's go over it again from the beginning.

The drip edge goes on first along the lower edge of the sheathing. If you are using ice shield, unroll it into place along the lower edge of the sheathing. The ice shield should be stuck down to the sheathing and the long leg of the drip edge. Next, cover the ice shield with No. 15 felt and then felt in the entire section of roof you have torn off. Finally, lay the roof edge up the rakes with the long leg of the roof edge *over* the No. 15 felt. Then you can begin marking the roof to lay shingles.

This sequence is designed so that the roof edge will protect your sheathing and rake board from windblown rain. The long leg of the roof edge carries the water out on top of the felt, fulfilling its function as a backup roofing system. The water runs over and down the felt (under the new shingles). When the water hits the bottom of the roof, the felt carries it over the long leg of the drip edge. The water drops from the lower lip of the drip edge into the gutter and it's gone.

The No. 15 felt and drip edge serve in the same way if your roof loses a shingle in a severe storm or if shingles fail at the end of your roof's serviceable life.

8.
Felt

No. 15 felt is the modern designation for the old style of felt, which weighed 15 pounds per roofing square. In other words, under each 10' x 10' section of roof, there were 15 pounds of roofing felt. In the same way that a modern 2 x 4 doesn't still measure 2" x 4", No. 15 felt no longer weighs 15 pounds per square.

Roofing felt is impregnated with asphalt to make it waterproof. When it is laid properly on your roof, the felt acts as a backup roofing system in case your shingles fail. The felt also serves as a cushion and moisture shield between the sheathing and shingles. Roofing felt is also commonly referred to as "paper."

A standard roll of No. 15 felt is 36 inches wide and covers 4 squares of roof. When you order, make sure you specify that you want *lined* felt. The lines run the length of the roll and show various overlaps. The lines 2 inches from each edge of the roll mark the standard vertical overlap. There is a line along the center of the roll for use in low slope (less than 3/12 pitch) roofs where you need to double your courses of felt to give maximum protection.

The Building Officials and Code Administrators' (BOCA) National Building Code is referred to here as "the Building Code." The Building Code *doesn't* require the use of felt on roofs steeper than 7/12. If you are contracting for a roof steeper than 7/12, *insist* that your contractor felt it in.

SIMPLEX NAILS VS. STAPLES

If you were a building contractor and you were installing felt to keep your basic framing and flooring dry until you could do the roofing a few weeks later, you would install your felt with *simplex nails.* The term "simplex nails" is another example of a brand name that has become generic. Simplex nails are nails driven through a 1-inch diameter washer or a 1-inch square washer to hold the felt in place. The broad holding surface of the washers helps keep the wind from tearing the felt off.

A *swing tacker* is a stapler that you swing like a hammer in order to pound your staples. I used a swing tacker for felting most of the time. I was always coming back with the shingles the same day and once the shingles are nailed on, the simplex nails or 3/8-inch staples have done their job. You will probably discover it is cheaper to buy the tacker and two boxes of 3/8-inch staples than to buy a 50 pound box of simplex nails. The tacker is also many times faster than nailing simplex nails.

I kept a box of simplex nails handy and used them on steeper roofs where I was sure the felt would let go if I just stapled it. I also used the simplex nails along the horizontal overlaps in the felt if it looked like a storm was going to catch me. A box of simplex nails lasted me a long time.

For the home roofer's purposes I recommend the swing tacker. There are several good brands on the market.

INSTALLATION OF FELT

A straight blade works best when cutting felt. Cut the paper wrapping off the roll. Be careful not to slice into the felt. Carry the roll down to the bottom of the roof at the rake. Position the roll straight up and down on the sheathing. The loose end should unroll from the bottom of the roll and lie flat when you fasten the felt down. Scoot the roll over so that the end of the roll is even with the rake or face of the rake board. (You don't want the felt hanging out beneath the overhanging shingles.) Slide the roll down until the lower edge slightly ($1/4$ inch) overlaps the outer edge of the fascia board (or drip edge).

When the felt is in position, staple the center of it near the rake several times all within a 1-inch (or so) diameter. These staples should run vertically up and down the roof. A staple is strongest in a direction perpendicular to the length of the staple. You are going to be using this cluster of staples in the center of the roll to pull against to adjust and tighten the roll. Slowly unroll the felt across the roof, keeping the bottom edge slightly over the fascia. When you have gone 7 to 10 feet, check and see if the felt has wandered up or down. You can raise the roll and pull slightly against the cluster of staples in the pivot point to shift the roll up or down. When you have the felt in position, tack it down with the swing tacker. All the staples, except the ones in the pivot point, should be horizontal across the roof. Why? Because the felt will try to tear away down the roof as you walk on it.

Roll out several more feet. The felt should stay in position or require only a slight adjustment. There shouldn't be any ripples in the felt. Again, staple it in position.

When you get near the far rake, staple the felt down securely 4 or 5 feet from the rake. Turn around to face this rake and roll the felt in front of you until you get almost to the rake. Pull the roll back toward you, backrolling a little extra felt off the roll. Stand on the felt gingerly so it won't slide down the sheathing, and raise the roll up above the edge of the rake. Keeping the felt tight on the roll, bump the bottom of the roll down soundly on the edge of the rake. Pull the roll down on the roof, backrolling the felt again until you can see the line the rake made in the felt. Use the straight blade to cut up this line. Take the roll up the roof and secure it against a plumbing vent, chimney, or something else solid. Smooth the cut end of the felt you laid on the sheathing and staple the end down tight. Bumping the roll down on the rake to mark where you cut *should* (with practice) give you a cut that is exactly even with the rake. Go back across the course of felt you just completed and staple it from top to bottom every few feet. (You don't want to mark the courses and then have the felt slip. You especially don't want it to slip as you're walking across it.)

Retrieve the roll of felt and start at the rake where you ended the first run of felt. Set the end of the felt even with the rake and overlapping down onto the 2-inch line on the first run of felt. Drive your pivot staples. Lay the second run of felt.

Stop stapling several feet from a plumbing vent and roll the felt over to it. Mark the location of the vent on the underside of the felt with the back of your straight blade. Backroll the felt until the area to be cut is suspended away from the roll, and cut out a rough hole for the plumbing vent. Raise the roll up and over the vent and settle the felt down around the vent. You may have to do some additional trimming to make sure there are no buckles or ripples in the felt. Staple the felt in position.

It's too easy to end up with a gaping hole torn around the plumbing vent. Finish the run of felt to the rake and staple all of it down solid except right around the vent. Let's say you have a 2-foot diameter hole torn around the vent. Cut 3 feet of felt off the end of the roll and bring the patch piece back to the plumbing vent. Rest the patch over the vent and cut out a circle for the vent. Slide the patch down over the vent until the patch is flush on top of the main felt. If you located the patch over the vent properly, you have a 6-inch horizontal overlap of the patch around the edges of the torn hole in the main felt. It's better (and easier) to have too big a patch than one that is too small. Continue the courses of felt on up the roof.

Murphy's Law dictates that you will never run out of felt exactly at the rake. You will always run out part of the way across the roof. Horizontal joints must have a 6-inch overlap. Overlap the end of the new roll 6 inches over the end of the old roll. Line the roll up horizontally and staple a pivot point for the new roll. Continue laying felt with the new roll.

When you get up to the ridge of the roof, lap the felt over the ridge to the other side. This lap or a second lap over the ridge must cover the top of the old shingles you left in place on the other side of the roof. This seals the ridge in case of rain during the night. Weight down the felt over the ridge by laying bundles of shingles along the horizontal edge over the old roof.

If a stockpile of shingles is sitting on the ridge in your way and it is definitely going to rain that night, use up the shingles you need to lay the side you tore off. You only have to move those shingles once. At the end of the day move the rest of your stockpile down onto the new shingle roof. (You don't want the stockpile in the way again when you tear off the opposite side of the roof.) Lay the bundles down gently on the new shingles so you don't scar the new roof. Now tear off and "felt in" the ridge where the stockpile was. Weight the felt down well with bundles, and the ridge is sealed.

If the wind picks up while you are "felting in" the roof, shorten the distance you go before you tack the felt in place. If you go as far as 10 feet on a windy day, the wind will get under the felt and lift it up, causing real problems.

In high wind or on an extremely steep roof, I unrolled the felt one run across the roof at a time and then shingled over that run, leaving room at the top of the felt for the 2-inch vertical joint when I overlapped the next run of felt.

If you get a bad ripple in the felt, it can show through the shingles. Slice down the length of the ripple with the straight blade. This will allow the ripple to lie flat. Cut a patch piece off the roll of felt, making sure the patch is big enough to cover the sliced ripple. Cut a horizontal slice (the same width as the patch piece) above the sliced ripple. Slide the patch piece 2 inches into the horizontal slice above the ripple. Staple the patch in place over the ripple. You have a 2-inch vertical overlap at the horizontal slice and should allow the patch to extend 6 inches on either side of the ends of the sliced ripple. By patching this way, you have maintained the 2-inch vertical joint and the 6-inch horizontal joints. The felt can continue to function as a backup roof.

Nail: Many roofers slice any ripples in their felt, but very few will patch over the ripple after they cut it.

CAUTIONS

1. You are working backward as you unroll and staple the felt. *Stay aware of any penetrations that could trip you and stay especially aware of the rake on the opposite end.* Though it seems impossible when you are roofing, you can get so engrossed in keeping the felt smooth that you forget where that other rake is. You can back off the roof much more easily than you would think. It won't be any consolation to know you weren't the first.

2. If the roll of felt starts rolling down the roof, don't try to stop it. It's heavy enough and will have enough speed to take you off with it. When a roll does go sailing, the felt usually ends up undamaged anyway. Don't dive or run down the roof for any reason unless it's to try to save someone else. Life is too sweet to risk it for a roll of felt.

3. Keep everyone out of the area below where you are working. Virtually anything falling off the roof will have enough speed when it falls to hit with killing force. This is as obvious as not backing off the rakes, but guess where people naturally stand to talk to someone on the roof or watch the work? One of my biggest problems was with little boys who were fascinated with the activity. Don't let anyone put himself in danger, and you won't be forced to dive for something to protect someone below.

4. Felt can be secure all morning, and you can be walking on it with confidence, when all of sudden it lets go. The sun heats up the black felt and the asphalt component gets hotter and hotter until the felt reacts like a greasy rag under your feet. You will develop a feel for how many staples are needed to hold the felt on the sheathing. Remember, staples are cheap, and it's better to have too many than too few.

5. Each roll of felt carries the warning that it is not to be used as a roof. Contractors on new construction leave a roof felted in for weeks before the shingles are laid, but this is not the intended use of the felt. In an emergency, felt will turn water and can save your home. However, the wind can remove the felt even if it is weighted down. Water can blow under the felt despite your best efforts, or the felt can tear. Rely on felt only in an emergency when you have no other choice.

John Zarbo and Matt Porter laying Number 15 felt. Matt is holding pressure on the roll to keep the felt tight. The swing tacker is many times cheaper and quicker than simplex nails.

We tore off both sides of the upper roof and papered it in. Naturally, thunderstorms closed in on us, so we weighted down the rakes and horizontal laps with unbroken bundles. No leaks! Doug Merold is helping lay the starter and first courses.

9.
Shingling a Tear-Off

Chapter 5 shows you how to overlay the original roof. An overlay requires less labor, involves less risk, and generates less scrap material to be hauled and disposed of than a complete tear-off. Overlays are covered early in this book because you should check that option thoroughly before deciding you have to tear off.

We covered the setting of verticals (vertical lines — base line and offset line) and the line for the starter course in Chapter 5. Let's run through the setup briefly again, using a different width roof from the 35'8" roof used in Chapter 5.

If you grasped every word in Chapter 5, you will find some of this chapter redundant. But don't worry, there is some good new information mixed in here, too.

You measure the roof and it is 37'6" wide. 37'6"/3' per shingle = 12 1/2 shingles. You want the lines near mid-roof, or about 18 feet from the rake. Pull from the *left* rake. We want 2 inches of shingle overhanging the rake so we can cut 1 inch off and get a perfectly straight cut when we trim the rake.

Your base line would be 18' - 2" = 17'10"
Your offset line would be 17'10" - 6" = 17'4"

Let me show you what size tabs you will end up with at our right rake.

Base line: 37'6" - 17'10" = 36'18" - 17'10" = 19'8"
Offset line: 37'6" - 17'4" = 36'18" - 17'4" = 20'2"

Both of these measurements are to the face of the right rake. Remember, we want to have a 1-inch overhang, so add an inch to each measurement to get 19'9" and 20'3". Shingles have a key cut out every foot, so at the right rake you will have a 9-inch tab and a 3-inch tab. (You will have a lot of shingle, 1'4" and 10", sticking out over the right rake. Rough-trim the shingles before you nail them in place. That much length will drop down and break the shingle right away if you don't rough-trim it first.)

The reason for checking the tab sizes at the rake is that you don't want to end up with a narrow sliver of a tab overhanging the rakes. If you have a 1-inch tab at the far rake, the key is directly above the rake edge, and the 1-inch tab has no strength. The 1-inch tab will gradually weaken and droop over the edge. I call these "pigtail tabs" because they end up looking like a line of pigtails drooping down over the rake board.

Nail: Most roofers pull in 36" - 2" = 34" and 34" - 6" = 28" for their base and offset lines. They let the tabs along the right rake end up whatever size they happen to be. That is the reason for so many homes with the weak, dangling "pigtail tabs."

The shingle layout will not be centered on the roof if we lay it out as we computed it. To make it even, the tabs on the left should lose another inch, and the tabs on the right should gain an inch. In effect, we need to shift the whole shingle layout 1 inch to the left. This is easy to do. Just subtract another inch from the initial measurements you made for the baseline and offset lines:

Pull over 18' - 2" - 1" = 17'9" for the base line
Pull over 17'9" - 6" = 17'3" for the offset line

Measure over 17'9" from the left rake along the ridge of the roof. Drive a nail at this measurement near the ridge. This is the base line nail. Now drive another nail at 17'3". This is the offset line nail. Hook the chalk line to the baseline nail and gently run the line out down the roof. Try to keep the chalk line out of the area where the finished lines will be. "Ghost" lines where the chalk line bumped the felt can cause you to make errors later. At the bottom of the roof, measure over 17'9" and mark the felt with a vertical arrowhead, using the blunt back of your straight blade or a nail. Mark the 17'3" measurement too. Pop the chalk line at the 17'9" mark, and you have the base line. Rehook the line to the offset nail at the top of the roof and pull it taut over the 17'3" mark at the bottom of the roof. Pop it and you have the offset line.

HINT ON VERTICALS

If you have left shingles stocked at the ridge of the roof, chances are good that Murphy's Law has come into play again, and the stockpile is right where you want to put your verticals. You want to pop the verticals all the way from the fascia to the ridge. Just vary the verticals 3 feet one way or the other. Instead of 17'9" and 17'3" use 14'9" and 14'3" or 20'9" and 20'3" to get to one side or the other of the stockpile. Then you can pop the verticals all the way to the ridge.

HORIZONTAL COURSES

Now we need to set up our horizontal courses, and we are ready to start nailing shingles. You want the shingles to overhang the fascia board by between $1^1/_4$ inches and $1^1/_2$ inches, so the water will run off the end of the shingle to the ground or into the gutter. You want the horizontal chalk lines to mark the tops of the shingles. The shingles are 12 inches

high so 12" - $1^1/_2$" = $10^1/_2$". Measure from the front of the fascia board up the roof $10^1/_2$ inches and drive a roofing nail. Hook the chalk line. Go to the far end of the roof and mark the felt with a horizontal arrow $10^1/_2$ inches above the fascia. Pop a line. You now have the line for the top of the starter course.

Some styles of aluminum drip edge can extend 1 inch or more beyond the face of the fascia board. Extend the bottom edge of the shingles 1 inch beyond the outer lip of the drip edge. You don't want the bottom edges of the shingles to extend more than 2 inches into the gutter and start blocking it. If you are using drip edge, measure up the roof 11 inches (12" - 1" = 11") from the outer lip of the drip edge. Pop the line for the starter course.

Open a bundle of shingles. There are a few ways you can make it easier to use the shingles. Bring the bundle down to the lower center area of the roof and hold one end while you drop the bottom of the other end of the bundle on the roof. Remember, you have self-sealing shingles: they tend to stick together in the bundle even with the protective tape. Dropping one end is called "breaking the bundle" because the impact causes the shingles to shift and separate from each other while they are still wrapped in the bundle. Pull the paper wrapping apart at its seam in the back and discard the paper. (I have watched some roofers shred the wrapper into five or six small pieces. Then they have a bunch of small wind-blown scraps to pick up.)

Some people with super-flex knees can squat and duck-walk along the lower edge of the roof and be comfortable. My knees won't take that kind of abuse. I'm right-handed, so I'm more comfortable sitting with my right leg tucked up under me and my left foot down the roof as a brace and brake. Sitting is a little slower, but I can still walk the next morning.

Take a shingle and turn it so that the top edge overhangs the fascia and the bottom of the tabs is right on the line for the starter course. Slide it over until the end (not the side of the half-key cut) is centered on the base line. You are going to lay the starter course this way with the grit to the sky, the tabs facing up the roof, and the top edge overhanging the fascia. Nail this starter shingle in place, driving the nails slightly below the self-sealing strip.

Now lay the next shingle over toward the left rake of the roof. Butt it in against the end of the first shingle you laid. Butt it reasonably snug, but don't jam it. The shingles will expand and contract. If they are jammed tight, they will tend to buckle. This second starter shingle should also have its tabs heading up the roof with its top edge dangling over the fascia.

Now lay the first shingle of the first course right side up and directly over the starter course. The end of this shingle should be in perfect line with the offset line, which has been covered over by the first starter course shingle. The top of this shingle can be set even with the bottom of the tabs of the starter course. The tabs on the shingles of the first course should be over the solid edge of the starter course. (As we discussed in Chapter 5, let the first course drop down $1/4$ inch on the starter course.) Nail this first course shingle in place. The nails go in the shingle above the keys and just below the self-sealing strip. There has to be a nail at each end of the shingle and a nail above each of the keys, a total of four nails.

The upside-down starter course provides a straight, unbroken, strong edge beneath the tabs of the first course. Water running down the keys of the first course is carried on over the edge by the upside-down shingles of the starter course. This method gives the maximum protection against damage to the fascia board and the ends of the joists covered by the fascia board.

When you drive your nails just below the self-sealing strip, it forces the strip to bulge slightly, so the shingle in the course above it can adhere firmly to it. If you nail right in the self-sealing strip, you pull the strip down and away from the shingle above it. (Plus, you end up with asphalt all over the head of your hammer. When asphalt does accumulate on the head of the hammer, rub the head vigorously on the top portion of a shingle. Shingles are like giant sheets of sandpaper.)

You will note as you lay shingles that nailing just below the strip catches the top of the course of shingles below. This means that each shingle is held in place by not just four nails, but eight nails. This makes for a very strong roof.

A nail gun doesn't lend itself to the exactness of hand nailing. Just shoot the nails into the self-sealing strip. If you try to stay below the strip, you are going to end up with some exposed nails on the new roof. You will also end up damaging, and removing, a lot of shingles. If you gun your nails in the self-sealing strip you will still be nailing through the top of the shingles in the course below.

Nail: I stressed four nails per shingle because that is the way the roof is designed, and that is what the building code calls for. Some unscrupulous roofers cut the corner by three-nailing or only two-nailing a shingle.

The code calls for six staples per shingle. Each end of a shingle should have a staple. There should be a staple on either side of the two full keys for a total of six staples. I have never, ever found a roofing company that six-stapled shingles. The most they do is four and not always that.

Some companies will fire one staple above each of the two full keys in the shingle. They will then fasten one staple across the butted joints of the shingles. In other words, one leg of the staple is in one shingle and the other leg of the same staple is in the next shingle. The company may try to convince you this is four-stapling the shingle. Sorry, this is three-stapling the shingles, and it gives you half the fastening strength the code calls for!

Nail: Shorter staples are cheaper than longer staples, so guess which staples an uncaring roofer is going to buy. A short staple will penetrate the shingle and bite into the wood sheathing if your roof is a tear-off. In the case of an *overlay*, the shortest staples will just barely bite into the wood. In other words, the legs of the short staples don't penetrate the old shingles and all the way through the sheathing. The new shingles are not fastened thoroughly to the wood sheathing, which is where the holding power is. If your roofer uses the shortest staples, he has, in effect, fastened the shingles of your new roof to the shingles of the old roof.

Nail: Short staples have a nasty habit of penetrating part of the length of their leg and then bending over and lying on top of the shingle. Staples like this might as well not be there.

Nail: The final problem with staples is that they give maximum strength or have maximum holding power in the direction perpendicular to the staple itself. On a roof, you want your maximum strength to be in a vertical direction. (Gravity is trying to pull your shingles down the roof, and high winds will try to tear the shingles up the roof.) This means that you want the staples to be horizontal; you want the exposed top of the staple to be running across the roof, parallel to the fascia board or gutter.

As roofers work, they are swinging the staple gun in an arc. The staple fired at the far reach of that arc is not horizontal across the roof. Depending on the roofer, that staple may be almost vertical (or perpendicular to the fascia). A vertical staple gives the roof the least strength possible, even if the staple legs fully penetrate the sheathing.

These tricks give you some insight into the bad rap on staples. Six-stapling shingles with long staples will generally work out fine.

It's easy to get confused on the starter and first courses, so nail the upside-down starter course all the way across the roof to both rakes. Now come back and nail the first course directly over the starter course. Just remember to start this first course at the offset line.

Rough-trim the overhangs as you finish each course so the shingles don't sag down or break along the rake.

Measure the courses up the roof in 5-inch increments above the starter course. Lay the end of the steel tape against the top of the starter course, and mark arrows up the roof in 5-inch increments. Do

the same marking at the other rake. Drive nails in each mark up one of the rakes. It's tedious, but keep hooking the chalk line one nail at a time and pop the 5-inch course lines across the roof. After popping three or four lines you will start running out of chalk, and the lines will get faint. Reel the line back into the chalk box, then hook the nail clip. Unreel the line, banging the side of the box on the heel of your hand and keeping the box nose down as the line reels out. This will keep the chalk loose and get the maximum amount on the line. Popping lines on horizontal courses is one of those times when another set of hands is most welcome.

Nail: Most roofers don't mark every course. The better roofers will pop a line for every other course. A roofer may have a good feel for how far down the shingle he is laying should overlap the next course down the roof. He can keep the courses straight by popping a horizontal course line every 10 inches.

Unfortunately, too many roofers pop a line every five courses (25 inches) or even as much as every eight courses (40 inches). They take great pride in the time they shave off the job by taking such large skips. Unfortunately, their courses get increasingly wavy and often look like snakes before they get up to their next marked course, which is perfectly straight.

Your starter course began at the base line and the first course began even with the offset line. So the second course begins back at the base line again.

Lay the second course. Double nail the rakes. When you come to the rakes, nail the shingle in its normal place. Then nail the shingle again just inside the face of the rake board. *Keep the nail far enough in from the face of the rake board that you don't end up splitting the wood on the face of the rake board.* Nail a second nail straight up the shingle from the first rake nail. Be aware of where the key of the next shingle up the roof will fall and keep nails away from that key. Double nailing gives an extremely strong edge to the roof and will help keep the roof on during very high winds. Keep laying courses until you complete the fifth course.

Keep checking to make sure you alternate each course. Starting two consecutive courses on the same vertical line is an easy mistake to make: you don't want to lay half the roof before you discover it.

Now you can stock the roof. Let's say this section of roof is 18 feet high. You computed the area to be:

37.5' x 18' = 675 sq. ft. = 6.75 squares
6.75 sq. x 3 bundles/sq. = 20.25 bundles

You have already laid three bundles in the first five courses, so you need to stock 20 - 3 = 17 more bundles.

Break the bundles as you drop them and unwrap the bundles you stock on the lower third of the roof. Don't get in the habit of unwrapping bundles until you are ready to use them. Remember that sudden storms or winds can blow the shingles from the roof. You can place containers of regular roofing nails above the bundles, or you can spot a coil of nails for the nail gun above each bundle.

The patterning on the shingles is designed to be random if you lay the shingles across the roof as described. This is particularly true if you are drawing from different bundles going across a stocked roof.

Now you have five solidly nailed non-skid courses of shingles to work from. The edge may make you extremely nervous, but it's time to turn your back to the world and work up the roof. (If five courses still places you too close to the edge, lay two more courses, then turn and work up the roof. Don't worry, you'll get used to it and will probably end up turning after only four courses.) I found that standing and bending over using the gun was the easiest and quickest way for me to nail. (A gun is nice but too expensive to buy for a one-time effort. You also have to be extremely careful not to step on the hose to the gun. The hose will roll out from under you and throw you.) Most people who hand-nail either sit and scoot or work from the balls of their feet and their knees. It's exciting to look back and see the progress you are making. Keep checking to make sure you have alternated every course.

REINFORCING RAKES

Some roofers reinforce their rakes with a shingle running lengthwise up the rakes. It's like a starter course for rakes. Lay the starter course along the fascia. When you get to the rake, turn a shingle so that its top is 3/4 inch out over the face of the rake. The top of this shingle supports the protective overhang out over the rake board. Pop a chalk line up the rake 11 1/4 inches from the face of the rake. This is the line for the bottom of the tabs of the shingles. Lay the horizontal courses right over this rake shingle, allowing a 1-inch overhang as computed before. If you are using twenty-five-year shingles, let the rake shingle extend 1 inch over the face of the rake and let the course shingles overhang by 1 1/4 inches.

The rake shingle is a nice idea, especially with a twenty-year shingle. However, I encourage you to pay three or four dollars a square more and go with a twenty-five-year shingle instead of the twenty-year. A 1-inch overhang on a twenty-five-year roof shouldn't droop down, even without the rake shingle.

ELIMINATING NARROW SHINGLE PIECES; MOVING BUTTED JOINTS

It is not unusual to need a small portion of a tab of a shingle to complete a course. Often you only need a few more inches of shingle to complete a course at a valley, wall, chimney, or other obstruction. When this happens, don't just put in the small piece and nail it down. The small piece will be very weak and prone to blow off. In addition, you don't want to drive nails through a small piece near the "V" of the metal valley.

You don't want a small piece with its *open*, butted joint running close to the step flashing down a wall, chimney, or other feature on the roof. The butted joint could allow water to get in under the shingles.

The last shingle before you would have needed the small piece has three full tabs. Nail the first two nails in the shingle and sit above the shingle using your foot to hold the loose end down. Start the hook blade at the *center* of the last full keyway and cut up the shingle, angling the cut slightly back toward the center of the shingle. Remove and save the tab you just cut off and lay a whole new shingle in its place. Nail the replacement whole shingle as appropriate. Now when you make the final trim cut on the replacement shingle, you won't have a narrow, weak tab (where it can be blown off) or a joint in a leak-prone location.

Theoretically, when you cut off that last tab of the final whole shingle in the course, you should make the cut straight up from the center of the keyway. I found I was usually off a little and had to trim the top of the shingle more to butt the top of my replacement shingle in, and to have the proper width on the key. By starting at the center of the key and angling the cut back *very slightly* toward the center of the shingle, you will eliminate a lot of lost time spent trimming. The end of the replacement shingle will butt right against the center of the key you cut, and the resulting key will be the perfect width.

FELTING AND SHINGLING ON A STEEP ROOF OR WINDY DAY

Don't try to felt in the entire roof if the wind is fighting you. Staple down one run of felt at a time. Start the first run of felt above the fascia. Roll out a short length (5 to 7 feet) and staple it in place. Weight it down with bundles of shingles. Roll out and staple another short length. Weight it down and continue toward the far rake. Mark the starter course. Mark the verticals in the felt at the bottom of the roof near the fascia. Mark the verticals at the ridge and as you did before, drive nails part of the way down at each of these marks. Hook the nail hook on the nails and pop vertical chalk lines down the roof. It will look funny to only have verticals on one run of felt, but don't let it bother you.

Lay the starter course and first course of shingles. These will hold the first run of felt in place. Move the bundles up so they just cover the top of the felt and mark and pop the lines for the next four courses. Lay the four courses of shingles. Move the bundles out of the way on up the roof and come across with the second run of felt, stapling and weighting it

down as before. Now hook the chalk line on the nails at the ridge and hold the bottom of the chalk line *exactly* on the vertical lines on the top of the first run of felt. Pop the lines. Go on up with the next courses of shingles. Keep repeating this process one run of felt at a time until the reach the ridge. It's extra work to keep moving the bundles up and down, but it's more work to have the felt blow off after you just finished marking all the courses.

VERTICALS ON A STEEP ROOF

On an extremely steep roof, hook a chalk line, and unreel the amount of free line you will need to reach down to the run of felt you want to mark. Slide the chalk box down the roof, keeping it a few feet away from where the verticals will be, so you don't get "ghost" lines. Go back down the roof and pop the line.

Get back up on the ridge and repeat the process for the next vertical. This is another case where a second set of hands is welcome. Single-handed popping of verticals on steep roofs is time consuming (and strenuous), but it can be done. It seemed that when we came to a really mean, steep roof, no matter how good the pay was, my crew got sick or disappeared.

LEAVING THE ENDS LOOSE

Occasionally you have to stop laying the horizontal courses of shingles before you can finish the courses all the way across the roof. At the same time, you need to continue the courses on up the roof. When this happens, don't nail the end of the shingles in the courses you had to stop. By eliminating the fourth nail, you have "left the ends loose." You can raise the loose ends up later and slide new shingles under them to continue the course on across the roof. After you have tied the course in, you go back and nail down the loose end.

You can leave the ends loose by just leaving out the last nail in either the base or offset courses, whichever extends 6 inches out in the direction the course is heading. It's easy to lose track of where you are and mess this up. I just left the end loose on every course. Then I knew I had to nail the loose end of each shingle when I came back and continued the courses on across the roof.

NAILING HIGH

You may have to carry the course over an obstruction, leaving a gap in the courses on the other side of the obstruction. Nail the shingles in this kind of course at the very top of the shingle, keeping the nails on a vertical line directly above the keys. This is called "nailing the shingles high." The nails go in their normal position when you lay the courses above the course you "nailed high."

Nailing high allows you eventually to tie in the next course down the roof by sliding the shingles for that course in under the course you carried across the top of the obstruction. The nails on the shingle you slide in will be in their normal positions on the shingle. Then renail the course of shingles you "nailed high," placing the nails in their normal position also. (Don't worry about the nails you left at the top. They already did their job, and they aren't hurting anything.)

"Leaving the end loose" and "nailing high" are both ways of allowing you to come back to tie shingles in later. They are often done in conjunction as you will see in later chapters.

EMERGENCY CHOICES

Shingling the roof should go without any major problems. The worst that should happen is you may have to double-check yourself by taking another look at a chapter in this book. However, sudden changes in the weather may force you to get everything as watertight as possible and then get off the roof. There are things you can do if you keep an eye on the sky and have a little warning before a storm hits.

Let's suppose you have just finished felting in the roof and those towering, fluffy white, cumulonimbus clouds off to the west start to bunch together and turn gray. The section of roof has one plumbing vent

and a wall rising above one end of the roof. Stock the roof, spotting the bundles along the overlaps of the felt. On a 6/12 roof, lock the bundles in place (by laying a second bundle on top, with one edge on the roof and the other edge on top of the first bundle). Leave all the bundles wrapped, and don't bother to break the bundles by dropping one end first. Weight the felt down at the ridge. Get light stuff off the roof or weight it down with a bundle. Get any trash off the roof.

Next, shape the all-lead flashing and bed the base of it in place in a bead of roofing cement, which you have caulked on top of the felt and around the plumbing vent. Tuck the top of the flashing in the vent pipe.

The storm is forming, so you still have some time. Figure the verticals on the roof so that they run close (one or two shingles away) to the wall you need to flash. Mark and pop the verticals. Mark the starter course and then mark the first several of the horizontal courses. Nail the starter and first course shingles from the verticals to the wall, but leave the ends of the shingles loose at the verticals. Install the lower step flashing. Lay as many of the rest of the partial courses from the verticals to the wall as you can; leave the shingles loose at the verticals. Tie the step flashing in as you do each course. Water accumulates as it runs down a roof, so every additional 5-inch piece of step flashing tied into the end of its partial course of shingles is critical.

You may see that you can't get the wall completely shingled and step-flashed. Cut a length of felt off the roll and lay it up the wall, preferably getting the edge under the siding. Use bundles to weight the felt down in position up the wall.

Another option is to run a bead of roofing cement to tie the end of the felt into a brick or mortar wall. The step flashing will cover this, but it can be a mess when you come back later to finish the step flashing. It's worth risking the mess to keep the water out.

When the storm is definitely coming but it's several miles away, get off the roof. Lightning will strike several miles in advance of the actual rain. Plus, wet shingles are slick, and wet felt is extremely slick. You did the best you could, so just hope it won't leak. I have been forced to carry more than one plain felt roof through a sudden totally unpredicted storm.

Measures like these work very well. It helps to take a "what if" approach to roofing, then you don't panic and flounder if the horizon suddenly gets dark.

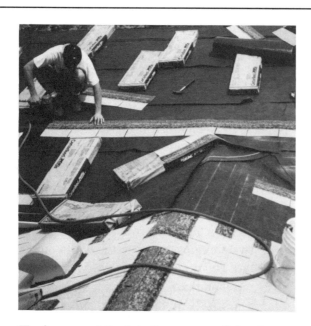

The lower roof ties into the upper roof, forming two valleys. The lower roof is an obstruction. I left the ends loose on the shingles to the right, and Doug is carrying the course over the ridge of the lower roof, "nailing high" as he goes.

Rain is coming. (1) Cut off the top of the cast iron plumbing vent. (2) Shape the all-lead flashing to the roof and try it on over the vent pipe. (3) Run a circle of mastic on the felt around the vent pipe. (4) Slide the unit down over the pipe, bedding the base in the mastic. (5) Nail the top corners of the base. (6) Tuck the top of the all-lead flashing up and over into the cast iron vent pipe. (7) Sweep metal filings off the roof, if you have time.

No leaks.

Mastic the felt directly to the chimney bricks in an emergency — no leaks.

10.
Plumbing Vents

Unless your backyard has a tall, narrow building with a crescent moon cut in the top of the door, there are plumbing vents on the roof of your home. Each sink drain, tub, or shower in the home has a vertical U-trap that stays full of water. The water in the traps and the water in the toilet keep methane and other gases of decomposition in the sewer system from traveling back into your home. However, these gases have to go somewhere, so the home has vertical vent pipes that carry these gases through the roof to the open air.

Vent pipes expand and contract as the temperature changes, so these pipes need to fit loosely through the sheathing and shingles. The pipe also needs to have room to play inside the neoprene collar or all-lead flashing, depending on which one you use. With both types of flashing, make sure the pipe is basically centered in its hole through the sheathing. The neoprene collar will let the pipe shift and rise and fall. The all-lead collar will be loose enough to let it do the same. A word of caution: when you tuck the top of the all-lead flashing into the top of the pipe, don't try to pound the lead down tight on the top of the pipe. You might tear the lead, and a tight fit won't allow the vent pipe room to move vertically.

Lay the courses up the roof until you reach a vent pipe. The lead in the all-lead flashing is slightly stronger than butter, so handle the unit carefully as you lift it out of the box. Slide the unit over the vent pipe to make sure it's the right size. Carefully check the soldered joint between the base plate and the vertical lead pipe portion of the unit. Make sure the solder isn't cracked or torn anywhere. Chances are, the vertical portion of the flashing is crimped or dented on one side. Place the unit on the roof beside the vent pipe. Turn the "good" side down the roof where it will be seen from below.

Pick the unit back up and lightly hammer beneath the upper half of the base until the base sits flush on the roof and the upright portion is vertical to the roof. Now try to slide the unit back over the vent pipe. The flashing unit may not slide over the pipe, due to your hammering the base. Sight down the vent pipe to see where the unit is hanging up. Remove the unit and gently tap around the inside of the hole in the base to round the vertical portion back out into shape. Now slide the unit over the vent pipe. If it goes on freely, take it off and place it to one side until you need it again.

Lay the next course of shingles. The vent pipe will stop the shingle from sliding up to the horizontal course line, but slide the shingle up into position as far as it will go. You need to cut a "U" notch in the top of the shingle. Use your knife to make a mark on the shingle. Keep the marks just outside of imaginary lines even with each side of the vent pipe. (Remember to leave the pipe some play. You saw how the old shingles were cut, allowing room all the way around the pipe.) Pull the shingle away and place the top of the shingle on the course line just beyond the pipe. Sight across the shingle and mark the shingle horizontally between the two vertical marks. This horizontal mark should be just below an imaginary horizontal line touching the lowest point of the vent pipe. The two vertical marks and the horizontal mark are the outline for the sides and bottom of the "U" notch for the plumbing vent.

Use the hook blade to cut the notch into the top of the shingle. (Cut the notch over the felt. It would be a shame to slice the shingles on the new section of roof.) Position the top of the U-notched shingle on the course line with the U-notch around the vent pipe. You will probably have to trim the shingle a couple of times to make the notch loose around the vent pipe. Nail the shingle in place, nailing as usual just above the keys. (See Figure 5 for the notching and shingling sequence around the vent flashing.)

You want the lower edge of the base of the flashing to extend down onto the exposed tabs of this shingle below the vent pipe. This means that the bottom of the base plate will show on the finished roof. This overlap onto the top of the exposed tabs forces any water hitting the base plate to continue down and over the shingle roof. Slide the vent flashing down into place, and see if the bottom edge of the base plate is below the top of the keys in the shingle you just notched and laid. If not, you need to lay another course, notching the top of a second shingle. You won't need to notch the tops of more than two shingles.

Place the flashing down over the vent pipe. Does the vent pipe stick up above the top of the vertical lead pipe of the all-lead flashing? If so, you need to cut the vent pipe so that it is an inch shorter than the vertical lead pipe of the all-lead flashing. You need to be able to tuck the top of the vertical portion of the all-lead flashing down into the inside of the vent pipe itself. If the vent pipe is too long, remove the all-lead flashing and use a hacksaw to cut the vent pipe so that it is an inch shorter than the flashing unit. A PVC (polyvinyl chloride, or plastic) pipe is not too difficult to cut. If the pipe is cast iron, you can cut it, but it takes some effort with a hacksaw. Make sure you get the measurements right; you don't want to have to cut a cast iron pipe twice. (Because I cut a lot of cast iron pipes in the course of a year, I used a hand-held electric hacksaw.)

Nail: Some contractors install the all-lead flashing, but they break the tops off any cast iron vent pipes with a hammer. The cast iron is brittle and it shatters. Breaking it with a hammer leaves a jagged top that can puncture the bend where the all-lead flashing tucks back down into the vent pipe. A shoddy contractor will often drop the broken pieces down the vent pipe, partially blocking it. Hammering can also crack the vent pipe down into the house and let noxious gases seep into the attic.

High winds can blow rain back up under the base of the vent flashing and cause a leak around the vent pipe's hole in the sheathing. Seal under the base of the vent flashing, using the caulk gun to run a heavy ring of roofing cement completely around the vent pipe. Keep the ring well within the area the base of the new vent flashing will cover. (Don't put down such a heavy bead of roofing cement that it will ooze out from under the base plate when you seat the base of the vent flashing in it.) The ring will go over the

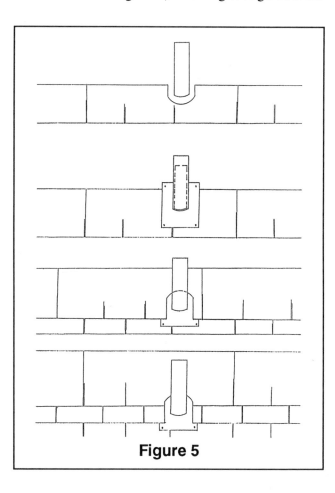

Figure 5

felt directly above the vent pipe, and it will also go over the new shingle surface beside and below the vent pipe. Slide the vent flashing down into position over the vent pipe and push the base down firmly, seating it in the ring of roofing cement.

Nail: Too many contractors do not bed the base plate of their vent flashing in roofing cement. The steeper the pitch of the roof, the stronger the wind must be to cause a severe leak. The flashing may never leak, or it may only leak inside the home during extremely severe rains. The damage will most likely appear when you roof the home again many years from now. When the roof is torn off in the far future, you will find that you have to replace sheathing and possibly repair a rafter. In exchange for that, these contractors save a few minutes and 25 cents worth of roofing cement.

Nail each corner of the base of the vent flashing down to the roof. Remember the problem with dissimilar metals. If you use a neoprene collar with a galvanized base, nail the base with a galvanized nail. If you used the all-lead collar, nail the base with an aluminum nail.

Nail: Too many roofers only nail the top corners of the base of the flashing. They will tell you that nailing the lower exposed corners can cause leaks and is unsightly. Of course, this is baloney. They are saving two nails and the time it takes to drive them and caulk over the heads. That unsecured lower edge of the flashing gives a high wind a place to start tearing at the roof.

Come across the roof with the next course of shingles. This time you will have to notch the bottom of the shingle. Mark the top and sides of the *inverted* "U" similar to the way you did the lower shingle(s). Mark the sides of the *inverted* "U" notch at the bottom edge of the exposed tab(s). Mark the top of the inverted "U" up on the shingle. Cut out the inverted "U" notch. Slide the shingle down into position with the notch up and over the lead pipe portion of the vent flashing.

Nail the upper shingle in place. If the top of the key is close to the vent pipe, either nail high at the top of

the shingle or nail off to the side of the key. Just don't drive the nail in its normal position over the key if that means the nail will add another penetration to the base of the flashing.

When you trim the shingle, leave aproximately 3/8-inch clearance around the vertical lead pipe of the flashing. Be sure to cut the legs of the notch straight down. These cuts are going to show on the finished roof. (When you trim the notch, be careful not to hook and cut the soft lead base plate under the shingle.)

Once this upper shingle is in position and has been trimmed, take out a little more insurance and run a bead of mastic back under the sides and top of the edges of the inverted "U" you cut in the tab of the upper shingle(s). Bed the shingle into the mastic by pressing the shingle firmly into the mastic.

Use the hammer to tuck the top of the vertical lead pipe gently into the inside of the PVC or cast iron vent pipe. Any rain that hits the top of the pipe will either run down the outside of the flashing and off the roof, or it will run down the inside of the vent pipe to the sewer system.

There is one more thing to do at the end of the job — caulking. Use a tube of aluminum-colored silicone caulk to seal the notched edge of the top shingle down to the base of the all-lead flashing. Also, caulk around the top half of the all-lead flashing to protect the soldered joint between the base and vertical section of the flashing. In effect, the 3/8-inch clearance around the top half of the vertical pipe of the vent flashing will be caulked-in solid. Use gutter seal (or aluminum caulk) to seal over the lower (exposed) nailheads. I waited until the end of the job to do the final surface caulking so the fresh caulk didn't get grit in it and I didn't drag my air lines through the fresh caulk.

By bedding the base of the flashing in roofing cement and tucking the top of the all-lead flashing down inside the vent pipe, you have sealed the unit. Your plumbing vent is not going to leak even if rain catches you the instant you finish installing the flashing and there is nothing on the roof but felt above the vent flashing. If the vent flashing is the all-lead type, it's not going to leak in the future either.

VENT FLASHING ON AN OVERLAY

The vent flashing on an overlay is done basically the same way a vent flashing is done on a tear-off. The main difference is that instead of using the chalk lines for the horizontal courses to lay the shingles, you are nesting the tops of the new shingles to the lower edges of the tabs of your old roof.

Wait until you get to each individual vent to remove the old vent flashing. You don't want to remove two or three vent flashings on a large section of roof and then have to worry about sealing them if you get caught in a sudden rainstorm.

Cut the old shingles overlaying the base of the old flashing. Keep the cuts just outside the edge of the buried base plate. Carefully remove the cut pieces of the old shingles to expose the base plate. *Save the pieces of shingle.* Pry the old flashing loose with the claw hammer; the nails should come loose with it. Slide the old flashing up and off the vent pipe.

Throw the old flashing away. You may be tempted to reuse an old all-lead flashing, but the solder between the base plate and vertical lead pipe may not last the life of the new roof. You're better off replacing the old all-lead flashing.

Fit and nail the cutout sections of old shingle back into their original position around the vent pipe. You don't want to leave these old pieces out, or you will cause a dip or swag under the new vent flashing.

Once you have the vent flashing removed and the old shingle pieces back in position, follow the procedure used earlier to vent flash on a tear-off. Just nest the shingles instead of following the lines.

Chapter 11 covers "Metal Chimneys, Pot Vents, and Power Ventilators." You will find that the method for weaving these into a shingle roof is basically the same as the method for a plumbing vent. The difference is one of scale.

Large obstructions stop the horizontal courses the same way a dormer, chimney, or skylight will. Chapter 12 covers "Resetting Base and Offset Lines Beyond Obstructions."

Course is up to the plumbing vent.

Notch the top of the next course. I put the notch I removed down on the white part of the shingle so you could see it better.

Circle around the plumbing vent with mastic. The upper part of the circle is on the felt, and the lower part of the circle is over the shingle.

Set all-lead flashing back over the vent pipe and nail it in all four corners (aluminum nails). The next course goes above the flashing unit, and the notch comes out of the bottom of this course. Don't nail the shingles through the base plate. Mastic above the vertical pipe of the all-lead flashing and partway down both sides. Bed this shingle in the mastic.

The next course over the top and I only had to take a slight notch out of the bottom of this one. I nailed this course in the normal spots because the nails are above the base plate of the flashing.

11.
Metal Chimneys, Pot Vents, and Power Ventilators

Metal chimneys, pot vents, and power ventilators are large roof penetrations. All of them have a vertical component that rises above the roof and a base woven into the shingles. The lower edge of all the base plates comes down over the tops of the exposed tabs of the lower course of shingles. Instead of a simple "U" notch, you cut several courses of shingles to fit along the sides of the vertical component. The bottoms of the tabs of the shingles laid around and over the vertical component are cut in a rounded curve to fit the outline of the top of the vertical component. (Chapter 12 shows you how to restore the base and offset line for the courses that are cut off by the obstructing unit.) Lay these courses back toward the obstruction from the new base and offset lines you reestablished beyond the obstruction. Cut the shingles to shape so they fit along the side of the obstruction. Figure 6 shows the completed installation of a power ventilator (dome removed).

METAL CHIMNEY ON A TEAR-OFF

A modern metal chimney consists of an interior pipe and an exterior pipe with an insulated gap between the two pipes. Even though the outer pipe

Figure 6

is insulated from the extreme heat of the inner pipe, the outer pipe still gets hot enough to burn you. There is a fixed flange that ties the metal chimney into the roof. The flange slopes down and away from the outer metal pipe before it flattens to form a base plate that ties into the shingles. The downward slope and additional distance from the outer pipe keep the base plate area of the flange cool so that it can safely be woven in with the shingles.

On a tear-off, remove all shingles and felt from around the base plate. Pull the nails in the base plate to remove the course(s) from underneath. When you lay the new felt, slide it under the base plate. Keep the felt well away from the wall of the metal chimney beneath the base plate. *If you jam the felt into the chimney under the base plate, you create a fire hazard.*

Lay the lower course by notching the top of the shingle for the metal chimney and sliding the shingle into position beneath the fixed base plate. Make sure you have notched the shingle so that there is plenty of space between the shingle and the metal chimney. When the base plate of the chimney flange covers the exposed portion of the lower course of shingles, try to run a bead of mastic under the base plate. This is tricky to do and may be impossible due to the strength of the metal and the fact that the flange is fixed. If you can't get a bead all the way around under the base, try to get a bead under the lower portion of the base plate. This will keep the rain from blowing back in under the lower edge.

Nail the base plate down to the lower course. I always drove the new nails through the old nail holes in the base plate. The nails go into the sheathing in the old holes, too, and won't hold like a nail in a new spot, but the flange is fixed in position on the chimney and, as you've discovered, it isn't going anywhere.

Cut and lay the shingles around the side and across the top. Pop the verticals for the interrupted courses and cut and lay the shingles back. It's fine to mastic the edges of the shingles down to the base and caulk the side and top of the cuts to the vertical wall of the flange. The lower part of the flange won't ever get hot enough to cause any problems.

METAL CHIMNEY ON AN OVERLAY

On an overlay, the strength of the fixed chimney flange will stop you from sliding another course of shingles beneath the base portion of the flange. Remove the nails from the exposed lower edge of the flange. Cut away the old course of shingle 2 inches out from the sides of the base plate. (You will have to cut part of the second course up along the sides too.) Remove the entire section(s) of shingle between the two cuts. Slide the old shingles out from under the base portion of the flange.

Lay a new course of shingles by sliding the new shingle beneath the base plate. Cut the top of the shingle so that it is well away from the metal chimney beneath the flange. You may have to slide the new shingle out a couple of times to trim it away from the metal chimney until it nests with the course of old shingles on both sides of the base.

Don't disturb the remaining old courses of shingle to the sides and across the top of the base. Overlay right on top of them. Keep the nails out of the base plate area and let the new shingles overlap the edges of the old shingles by $1/4$ inch. Mastic the new shingles to the old shingles and come back later and caulk the edges and the exposed nailheads at the bottom of the base.

DOMED VENTS AND POT VENTS

A domed ventilator depends on convection air to pull heat out of the attic. These ventilators consist of a base woven into the shingles and a vertical component, which opens above a hole cut in the sheathing. The weather is kept out by a protective dome over the vertical component. Most units also have insect screens. The smaller versions of domed vents are called "pot vents" because they look like upside-down pots on the roof.

GOOSENECKS

Some ventilators keep water out of the home by curving the sheet metal shaft up and over until the vent opening faces back down toward the surface of the shingles. This type is called a "gooseneck" for the obvious reason.

POWER VENTILATORS ON A TEAR-OFF

When a ventilator unit includes an electric motor and a fan, it becomes a "power ventilator." A power ventilator can best be described as a through-the-roof attic fan. Convection air currents carry a large volume of superheated air through a plain ventilator, but a power ventilator removes the heat more quickly.

The thermostat on most power ventilators is preset at a reasonable temperature. You can set the thermostat at the *attic* temperature you prefer to maintain. Remember, it's in an attic where temperatures easily soar over 120° F. Don't set the thermostat at 80° F., or the unit will run constantly during the day and into the night.

The basic steps of installing any ventilator are the same. However, if you are installing a new power ventilator you have a couple of extra steps. Remember the word ASS/U/ME, and don't weave a new ventilator into the shingles until you have tested it.

If the old power ventilator has been in the roof for ages, you should probably replace it while you are reroofing. If you replace the old power ventilator, make sure you buy a ventilator requiring the same size hole or one slightly *larger* than the unit you are replacing. Remove the old shingles and felt. Pry the old ventilator loose and unhook the connecting wires (with the power off). If the new power ventilator is larger than the old one, you may have to trim the sheathing back to match the size of the new opening. A saber saw is slow but does a good job of cutting a circular opening in the sheathing. Remember, there are rafters on either side of where you are cutting, so don't cut into them.

Now fit the framing and motor down into the hole through the sheathing and hook up the electric wires to the new ventilator. I recommend you get into the attic and hook up the electric wires as early as possible in the morning. People have been known to collapse in superheated attics, so don't wait until the hottest part of the day to hook up the ventilator.

Now let the power ventilator sit in position until you can see if the thermostat and fan motor are going to work properly. If you are the impatient type, you can run a drop light up into the attic and hold the bulb right under the thermostat until the heat makes the fan click on. Pull the light away: the thermostat should click the fan off fairly quickly. Repeat the process. If the fan clicks on and off properly, it's safe to go ahead and weave the power ventilator into the roof.

If you are installing a power ventilator in a new location, here are some basic ideas before you cut a hole in the roof. First, keep the ventilator(s) on the less visible portion (usually the back) of the roof. You don't want the power ventilator to become an outstanding architectural feature of your home.

Second, the ventilator should be located approximately a fourth of the way down the roof from the ridge. Heat rises and the highest heat is at the ridge, so why not put the ventilator right at the ridge? Any break in the shingles for any reason gives you a location for a potential leak. If you put the power ventilator right at the ridge of the roof, it is easier for rain whipping across the ridge to be driven up under the dome of the power ventilator and into your home. Also, the wind tends to have the greatest force and speed coming over the ridge. In a severe storm, a power ventilator is subject to the highest force at the ridge. The wind could tear the dome free from the unit and send it sailing.

Whether you need the power feature or not depends partially on the climate where you live. A power unit costs many times more than a domed ventilator or pot vent, and the motor is just one more stimulant for your electric meter. I sometimes think that knowing the fan is whirring away in the attic just makes people feel cooler. There are many choices of types of ventilators. The power ventilator is the only one that continues to cost money after you install it.

As with other roofing components, plastic power ventilator units are available, and they are cheaper than the metal ones. Stay away from plastic units. I have seen too many plastic domes that cracked or otherwise failed and allowed water into homes.

With the shingles and felt removed by a tear-off, you can locate the joists by spotting them through the cracks in the sheathing or seeing where the plywood is nailed up the rafters. You can cut out the circle of sheathing from on top of the roof.

You need to cut a hole through the shingles and sheathing and miss the rafters. You will need to know if the rafters are on 16-inch centers or 24-inch centers. In other words, what is the distance from the center of one rafter to the center of the next rafter in the attic? Power ventilators are made to fit between one of the two spacings. The package for the new ventilator usually includes a cardboard circle you can cut out and use as a pattern for cutting the necessary circle in the sheathing. This circle is the diameter of the segment of the base that will fit down into the sheathing. Cut this circle from its surrounding cardboard and take it and a drill into the attic with you. Select the location you want and hold the circle against the underside of the sheathing (or over the points of all the roofing nails). Drill four holes up and through the shingles on the outside to outline the diameter — top, bottom, and both sides. (Four nails driven through to the outside can accomplish the same task.)

You want the wood around the hole to be as strongly supported as possible. With plywood, try to center the circle near the vertical center of a full sheet of plywood. With planking, try to make sure all boards you cut extend across the adjacent rafter to the next rafter over. (You can normally tell by the color and grain of the wood if the plank is continuous on the other side of the rafter.) If you cut off a plank that ends at the rafter next to the cut, the plank is unsupported and will drop free if you stand on it. It may be necessary to box in around the power ventilator opening with 2 x 4s. Run the top and bottom 2 x 4s from rafter to rafter, and frame new side 2 x 4s between the top and bottom 2 x 4s. That way all of the plank sheathing is supported around the opening.

Take the circle back up on the roof with you and center it between the four holes you just drilled. Cut the old shingles and felt away using the circle pattern. Use a saber saw to cut the sheathing or go gently with a circular saw. Now set the power ventilator in place and go back to the attic and wire it. Make sure it works before proceeding. Lift the unit back out.

Shingle up one and perhaps two courses under the portion of the base that will be flush with the roof. You will have to trim the top of the shingles to keep the vent hole fully open. (A few blocked inches make a great deal of difference in the efficiency of the unit.) Now run a very heavy bead of mastic around the outside of the circle. Be careful you don't use so much around the bottom that it will ooze out from under the base. When you have the lower course or courses in place and the power ventilator bedded in a bead of mastic, nail every 6 inches across the base, sides, and top of the base plate (aluminum nails for an aluminum unit).

Usually four bolts (or nuts) hold the dome on a power ventilator. It is easier to remove the dome and trim and caulk the shingles in than it is to try to trim and work under the dome. After you remove the dome, screw the bolts back in place and weight the dome down out of the way on the roof. The wind can take the dome and send it sailing like a giant Frisbee. (Please don't ask how I know this. Just take my word for it.)

Lay the shingles up the sides. Don't nail any shingles through the base plate. You are going to mastic the ends of these shingles in place. The lower shingles covering the sides of the base plate are cut straight down from the widest part of the circular vertical component of the power ventilator. In other words, cut the lower side shingles straight down toward the gutter from the widest point on the "walls" of the power ventilator. These shingles will still cover the sides of the base plate by a few inches. You want the water running down the roof to run around the top half of the vertical component, then straight down from the sides and down the roof. (See Figure 6.) You can cut the shingles at the top half of the power ventilator so that they leave a 3/8-inch gap between the vertical component of the base and the cut edge of the shingle.

If the verticals fall so that a shingle would just be a weak sliver coming into the side of the power ventilator, cut a full tab off the last full shingle. Fit in a section of shingle that includes a full tab and the narrow piece you need. (This is just a reminder. We already discussed this method of adding strength.)

To pick up the base and offset lines on the other side of the power ventilator, lay the shingles loose right over the top and pop the verticals down (as shown in Chapter 12).

Once you have the power ventilator shingled in, lift the loose edges up the sides and across the top and run a bead of mastic under all the edges. Press the shingles down into the mastic to get them bedded. *Immediately*, run another bead of caulk along the cut edges on the sides and across the top of the base. Bolt the dome into its proper position before proceeding to lay shingles up the roof.

The course of shingles that first overlaps the sides will have its bottom edge resting out a few inches up from the bottom edge of the base plate. Don't mastic these bottom edges. If water does somehow penetrate around the ventilator, you want it to run beneath the side shingles and surface at the bottom of the base plate and from there to run down the roof on top of the new shingles. Mastic on the lower edges of shingles tends to hold the water beneath the shingle, so you don't want it there.

An 8-inch pot vent in place. The pot vent is bedded down in mastic, and all four corners are nailed. I papered over the ridge and caught the top of the vent for good measure.

12.
Resetting Base and Offset Lines Beyond Obstructions

A metal chimney, power ventilator, skylight, brick chimney, dormer, etc., will cut the courses off from the original base and offset lines. The courses below the obstruction continue across the roof uninterrupted to the far rake. You want the keyways of the courses broken by the obstruction to line up exactly with the keyways of the continuous courses. Let me show you two obstruction situations and how to establish new verticals in each case.

POWER VENTILATOR

Look back at the power ventilator in Figure 6. Your base line and offset line are to the left side. The size of the unit breaks the continuity of six courses. Lay the six courses up the left side of the unit and cut and trim them as shown. Carry the seventh course on across the top of the unit, nesting the tops of the new shingles to old shingles. Once you get to the top of the unit, nail the shingles in course 7 high. ("Nail the course high" by driving the nails at the top of the shingle, keeping the nails in a line directly above the keys.) Carry course 7 on beyond the power ventilator two (or three or four) shingles and stop laying in that course. When you stop laying in that course

"leave the end loose." ("Leaving the end loose" means you don't drive the fourth nail at the end of the shingle.)

Now bring the eighth, ninth, tenth, and eleventh courses on across, stopping them when you have laid the same number of shingles you laid in the seventh course. Nail courses 8, 9, 10, and 11 in their normal positions above the keys, but leave the ends loose on each of these courses.

Drive a nail directly above the end of the last shingle on course 11. Hook a chalk line to the nail and run the line down across the gap the power ventilator caused in the courses. Hold the chalk line down several courses into the courses that continued uninterrupted below the power ventilator. Make sure it lines up with the ends of courses 9 and 7 above the power ventilator. Pull the line tight and pop it. You have the new base line. Figure 7 shows the new base line established from the end of course 11.

Now lay the same number of shingles for course 12 as you did for the others coming across the top of the power ventilator. Nail directly above the end of the last shingle on course 12. Hook the chalk line, centering the line on the appropriate keyways on the

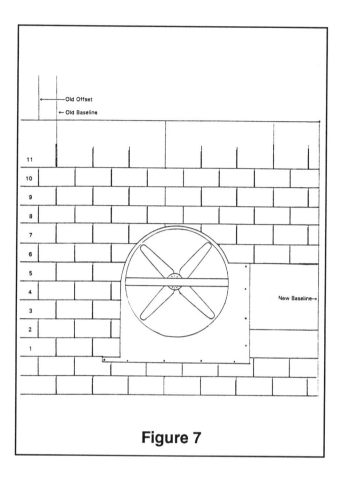

Figure 7

uninterrupted shingles (to hold on the keyways) gives you more accurate lines than just holding the line across one or two shingles at the top of the gap, and one or two shingles at the bottom of the gap. When you use this method of redoing the verticals, the keyways will turn out straight — as if there had never been a power ventilator in the way at all.

OBSTRUCTIONS AT THE RIDGE OF THE ROOF

An obstruction such as a brick chimney can reach to the ridge of the roof. In a case like that, lay uninterrupted courses across the bottom of the roof. Go several courses down from the last uninterrupted course you laid and find the end of a shingle in one course down from a base line course. Carefully set the nail hook under the full tab, making sure it is centered exactly below the joint where the two shingles in the base course above are butted together. Pull the chalk line gently straight up the roof and lay the line down. Now weight the hook down under the tab with a bundle of shingles and pull the chalk line directly up the roof. Pull it snug, centering it over the keyway of the several courses above where you have it hooked. Hold it up at the ridge and pop the new base line.

Set the nail hook up or down one tab, centering it directly below the butted joint of shingles in an offset course. Repeat the process and pop the offset line all the way to the ridge. Lay the courses back from the new verticals and tie the shingles into the obstruction. Then continue on across the roof.

If the bottom edge of a tab gets nicked when you pull the chalk line against it, push it down smooth with your finger or tap it smooth with a hammer. (Someone else holding the chalk line makes all this easier: he can hold the chalk line directly over the butted joint. I have described the method for setting the nail hook to pop a chalk line when you are working singlehanded. You will probably have to set and weight the nail hook more than once to get the chalk line to pull right over the joints.)

uninterrupted courses, and pop a new offset line across the gap. Lay the shingles back toward the power ventilator and trim them in.

You "nailed high" on course 7, so you can slide the shingles for course 6 directly under the shingles of course 7 and set them on the offset line to trim them into the power ventilator. Nail the shingles in course 6 normally. As you continue courses 7, 8, 9, 10, 11, and 12, you can butt the next shingle into its proper position because you "left the ends loose" on these courses.

Of course, if you are overlaying the old roof, you nest the new shingles to the old shingles to carry the horizontal courses. If a tear-off has a big obstruction, just measure and pop lines for 5-inch courses up both sides of the obstruction.

Going up as many courses as possible above the power ventilator before you nail above the end of the shingle, and going down several courses on the

13.
Step Flashing

Step flashing is metal laid with the shingles to seal along a straight vertical obstruction such as a wall, brick chimney, or skylight. All step flashing is laid basically the same way. You can buy step flashing in mill finish (unpainted) aluminum, enameled aluminum (black, brown, and white), copper, or galvanized metal (but don't use the galvanized).

Modern step flashing is manufactured from 5" x 7" rectangles. The rectangles are bent 90° so that a 2-inch leg goes up the wall and the 3-inch leg goes over the top of each shingle as you lay it. Always lay each piece of step flashing on each succeeding course at the same location on the shingle. Each piece of step flashing, or "step," is 7 inches long. The courses of shingle are only 5 inches. When you nail each succeeding piece of step at the same location on each succeeding shingle, the upper piece of step overlaps the lower piece by 2 inches. I always set my piece of step with the bottom of the step just above the self-sealing strip. Keeping the strip exposed lets the shingle seal all the way across to the vertical leg of the step.

If you have trouble visualizing this, lay down a shingle and rest a piece of step along the edge of the shingle with the bottom of the step just above the self-sealing strip. Place another shingle 5 inches up and rest another piece of step in the same location. Repeat the same process with a third and fourth shingle. Now grab the 2-inch vertical legs of the four pieces of step and pull them out together. The water is going downhill and all the step has a 2-inch downhill lap. It *can't* leak.

Nail: Some contractors don't lay their step as they lay each course. They leave the ends of the shingles loose along the wall and come back later to lay the step flashing. It's quicker for them to do the step all at once. The problem is, the roofer can get called away to help with something else, or take a coffee break, or just plain get careless. At any rate, it's easy to skip a piece of step this way. When a piece is left out, the wall has a 3-inch unprotected gap instead of an unbroken series of step flashing with 2-inch laps all the way down. You may get lucky and the felt will carry you for awhile, but eventually the felt will weaken, and it's going to leak. In a severe storm, it's going to leak a lot.

Some roofers nail their step high, thinking that will help keep it from leaking. This practice tends to raise the bottom of each individual piece of step, which tends to lift the tab of the shingle laid above it. This makes the shingles look ragged, and the loose edges are subject to wind damage.

I nailed my step near the bottom. The nailhead is down flush, and the metal in the step, shingle, top of the next lower step, and felt all seal around the shaft of the nail. It's not going to leak.

Nail: We have covered this once, but it is worth repeating. Too many contractors mix dissimilar metals. They use the light .019 aluminum and nail it with galvanized roofing nails. The galvanic action between the nails and pieces of step could leave a series of corroded holes in the entire length of the step flashing. If you have the contractor lay twenty-five- or thirty-year shingles, the step flashing may end up leaking like a trickler hose before the end of the useful life of the roof.

If a keyway hits close to the step flashing, you don't need to nail above that key. When you nail the step in place, you lock the shingle down, too.

Water comes around the edges of the shingles at the wall (or obstruction). The water gets on the 3-inch leg of step beneath the shingle. The 3-inch leg drops the water straight down the roof and out on top of the shingle to which the step is nailed. The water is now on top of the shingles and continues flowing down the roof on top of the shingles. If a little water runs around the side of another shingle farther down the roof, the step at that spot carries it down and diverts it back up on top of the shingles.

If by chance a little moisture runs off the 3-inch leg of a piece of step, the water flows over the shingle the step is nailed to until it runs over the exposed surface of the shingle and from there, down the roof. Avoid having a butt joint right at the step flashing. Cut off the last tab of the last whole shingle you lay. Then lay and trim a whole shingle so you have a full tab plus a part of a tab going into the wall.

If you do leave the joint near the step, the water will roll off the 3-inch leg of the step and down through the joint in the shingles to the felt. The felt might carry it for a while, but twenty-plus years is asking too much of the felt.

Nail: I can't tell you how many experienced roofers came to work for me and looked at me like I was crazy when I stopped them from putting a construction joint right beside the 3-inch leg of the step. They just never realized how step flashing really works. You know how they had been doing step flashing before.

The upper leg of the step flashing can go behind wood, aluminum, or vinyl siding. It can also go beneath a skirt flashing that is tied into the wall or side of the chimney. The skirt flashing is tied into the wall and caulked along its top edge. The siding or skirt flashing keeps any water from infiltrating the top of the step.

When a wall rises above an up-and-over roof, you need to step flash the ridge. I ran my pieces of step up each side until the top of the last piece of step on each side of the roof stopped right at the ridge. I then cut down the center of the 2-inch leg of a piece of step and bent the 3-inch leg in the middle to the shape of the ridge. I usually had to go to one side or the other of the up-and-over roof, and slide this piece up under the siding. Then I moved the piece up and bent the 3 inches over the step on the other side of the roof. This bent piece went over the straight pieces to continue the downhill lap from the ridge on down both sides of the roof.

I would then try to slide a small rectangular piece of metal in front of the cut that opened to a "V" notch when I bent the piece of step over the ridge. I caulked the notch, kneading the caulk back in behind the notch in the 2-inch side of the step. The caulk not only sealed the notch, but would hold the small rectangular patch in place in front of the V-notch.

Nail: I saw a lot of older roofs where the contractor had just run his step up both sides of the roof to the ridge and gunked the tops of the straight pieces together with mastic. The roofer had then jammed mastic back in against the wall too. It works and is OK until the mastic cracks. I saw a couple of roofs where the step flashing just stopped near the ridge, and there was no mastic or anything protecting the wall at the ridge. It's true that water accumulates down the roof, but the ridge has to have some protection. There was some damage to the sheathing at these unprotected ridges.

On low slope roofs, the shingles are reduced from 5-inch to 4-inch courses. This means the step automatically goes from a 2-inch to a 3-inch downhill overlap.

In addition, you can buy oversized pieces of step. The oversized pieces, such as the 9" x 12", cost more than the standard 5" x 7" step, but if you have a particularly trouble-prone location on your home, consider the larger sized step.

In a pinch, you can make step flashing yourself, but the manufactured step is cheap enough that it isn't worth your time.

14.
Tying into a Wall

An example of a roof tying into a wall occurs when a home has a one-story wing that ties into a two-story main house.

TYING IN A LEAN-TO ROOF

When a simple lean-to roof like you see on many screen porches ties into a wall, long lengths of single bend metal flashing are used. The metal should be fabricated in a metal brake. A metal brake is a machine that holds sheet metal fast and bends the metal over a straight edge. The length of the pieces of bent metal depends on the size, or working length, of the metal brake. I used an 8-foot brake, which means I could cut an 8-foot length of coil, clamp it down, and bend the entire length over a straightedge. A sheet metal shop may have brakes ranging up to 14 feet. I found that 8 feet was long enough for roofing.

If access to a brake is impossible, you can bend the metal with 2 x 4's. Lay the metal on a work bench with the dimple marks for the bend even with the edge of the bench. Place a straight 2 x 4 over the metal. Line the edge of the 2 x 4 up right over the marks (and the edge of the bench) and clamp the ends of the 2 x 4 down with large C-clamps. Place a second 2 x 4 under the metal at the edge of the bench and raise the second 2 x 4 up, rolling the 2 x 4 to keep the surface of the 2 x 4 flush on the bending metal. (Friends can be substituted for C-clamps.) Bending with 2 x 4's works, but the metal brake

gives a perfect bend, which adds to the finished appearance of your work. It's worth it to go to a sheet metal shop and form the metal on a brake.

Whether you use a brake or 2 x 4's, bend the metal to an angle slightly less than the angle you actually need. When the metal is nailed in place, it's natural spring will push it down toward the shingle surface and the wall. If you make a mistake and get too great an angle, the metal tries to spring back up to that angle and pull away from the shingles and the wall.

Lay the last course of shingle right up to the wall. The leg of the metal covering over the shingles provides the protection from wind-blown rain. Naturally, this leg down the roof must extend over the keys and down a little over the finished tabs. If you are working from a narrow metal coil or the leg up the wall has to be long, you may have to force another course of shingles into the wall by cutting the tops of the shingles. (If you have a choice on lengths of legs, make the leg down the roof as long as is reasonably possible.) When you must cut the top course, pop a horizontal course line across the tabs of the previous course. To cut a shingle, turn the shingle 180° so that the bottoms of the tabs touch the wall. Trim the top off the shingle, using the chalk line on top of the previous course of shingles as a guide. Use a hook blade to raise the shingle as you cut it; that way you won't cut into the new roof underneath.

Trim the rakes.

The lengths of bent metal flashing should overlap each other by 6 inches. For example, if your lean-to roof is 18 feet rake to rake, and you have access to an 8-foot metal brake, you will need three pieces of bent flashing. You would need 19 feet of metal (18'+ 6" + 6" = 19') to make the three pieces you need to flash this 18 feet. You should fabricate and install the metal so that the visible distance between the joints is the same. On an 8-foot brake, bend two pieces that are 6'6" long and one piece that is 6 feet long. Lay a 6'6" piece first, lap it with the second 6'6" piece, and lap over that with the 6-foot piece. From the ground it will have the symmetrical, visual effect of three 6-foot pieces of bent metal flashing butted together.

Pop a chalk line on the shingles to mark the lower edge of the bent flashing. (Your wall may bow in or out slightly, and you want the bottom edge of the metal to stay straight.) Slide the upper leg of the flashing up under the siding or skirt flashing of the wall. The end of the first piece of 6'6" flashing should be exactly even with the edge of shingle at the rake. Run a heavy bead of mastic under the lower leg of the flashing so rain can't blow underneath. Nail along the bottom edge of the lower leg. Locate each nail 1 inch above the bottom edge of the lower leg of metal and center the nails over each tab. When you mastic the lower edge of the metal and nail it every foot, it's not going anywhere.

Set the second 6'6" metal piece in place by sliding it up behind the siding or skirt flashing. Slide it until the end overlaps the end of the first piece by 6 inches. Run a bead of mastic under this overlap. This will keep rain from blowing into these horizontal joints.

If the top leg slides in behind the siding, you will have to determine what length of upper leg you want. The leg of the bent flashing you removed will give you a clue. You may have to notch the top of the upper leg around siding nails when you slide the new flashing up in behind the siding.

If you are nailing the top of metal flashing into a brick wall, you will find that mortar nails have the nasty habit of dancing around and augering a larger hole in the metal. Mortar nails can also split and throw shards of metal if you really smack them. I always pre-drilled nail holes right through the metal and mortar. I used a mortar bit slightly smaller than the shaft of my mortar nails. Drilling made driving the mortar nail easier, and the metal trim stayed in better shape.

Caulk the top edge of the upper leg to the brick. Make sure the caulk also covers the heads of the mortar nails. At the end of the job, smooth aluminum caulk or gutter seal over the heads of the nails along the lower edge of the bent metal flashing. When that is done, the lean-to roof is flashed into the wall for the life of the new roof.

TYING IN AN UP-AND-OVER (SADDLE) ROOF

Step flashing is used to tie an up-and-over roof into a wall. I always trimmed the 2-inch leg that went up the wall, so that the 3-inch leg would extend with the shingle the 1 1/2 inches over the fascia board. Cut up the 2-inch leg, keeping the cut parallel to and 3/8 to 1/2 inch from the bend in the metal. Place the 3-inch leg down on the felt and make a cut from the top of the 2-inch leg down to intercept the first cut you made. The angle of the second cut should roughly match the angle the corner of the siding makes with the surface of the roof. (See Figure 8.) Cutting a piece of step in this way will let you slide the 2-inch leg down behind the siding. A part of the 3-inch leg will stick out over the edge of the fascia. I didn't tell you how long the cut parallel to the bend should be, because this length will vary from house to house. You may have to trim this piece of step a couple of times to make it fit.

The bottom of this first piece of step should be even with the lower edge of the starter shingle. The first piece of step should also be nailed on top of the starter shingle. When you lay the shingle for the first course, put another piece of step on top of it and place the bottom edge right at the top of the keys. Immediately cover the piece of step over the starter course with another piece of step, but locate the bottom of this overlapping piece just above the self-

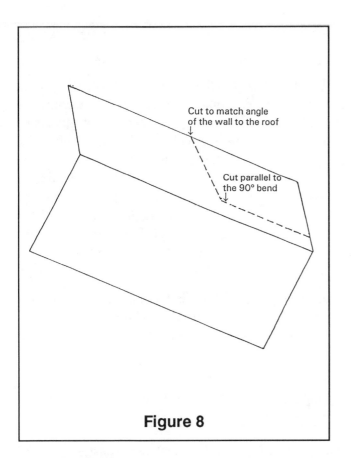

Cut to match angle
of the wall to the roof

Cut parallel to
the 90° bend

Figure 8

sealing strip on the starter shingle. By doubling up the step on top of the starter shingle, you will properly maintain the required 2-inch downhill overlap for the piece of step flashing that goes just above the self-sealing strip on the *second* course of shingles. The downhill overlap on all of the step flashing will be continuous from the ridge of the roof all the way down over the fascia.

"CUTTING IN" TO A WALL

It is extremely unlikely that any of the shingles will fit snugly against the vertical leg of the step flashing along the wall without being cut to the proper size. Let's assume you need to trim the final shingle on the starter course. Lay the final shingle in position as well as you can. Now flip the starter course shingle end over end. Put it back in its basic position and slide the end closest to the wall over until it is almost touching the wall. The tabs should still be pointing up the roof toward you: the grit side should be down toward the roof. Now locate the leading

edge of the last shingle you nailed in place on the starter course. Take the hook blade and cut the shingle you flipped even with the leading edge of the last shingle you nailed. The cut section, which is farthest from the wall, is "waste"; lay that piece aside. Flip the piece you need end over end again. The grit side should be up and the tabs should still be sticking up the roof. This piece should fit nicely in the gap you had in the starter course. To check yourself, notice that the starter course (and the rest of the courses) will continue with full 12-inch tabs all the way across the roof until you get to the final tab you had to cut to fit the last shingle into the wall. Nail the piece of step you trimmed so that its bottom is even with the bottom edge of the starter shingle.

(If you aren't getting rid of the scrap pieces right away, keep the self-sealing strips facing up and the tape side down on the roof. Otherwise, the sun will seal the trash to the new roof, and when you pull the trash loose it will leave a melted asphalt stain.)

Lay the first course (right-side-up course) to the wall. Again, the gap won't take a whole shingle. Spin the shingle 180 degrees. The grit side is facing up and the tabs are up the roof. Slide the end of the shingle into position against the wall and cut the shingle even with the leading edge of the last full shingle you nailed. Lay aside the cut piece farthest from the wall. Spin the shingle 180 degrees and slide it into the gap. It should fit properly, and you should have continuous 12-inch tabs right to the tab you cut. Nail this shingle in place.

Nail the doubled piece of step over this first course shingle as demonstrated. Cut in the second course and nail its piece of step in position with the bottom edge just above the self-sealing strip. Do the same up the roof with succeeding courses.

Nail: Because the base line and offset line are staggered 6 inches, some roofers cut the shingle pieces that butt in for the first and second course and use them as a pattern to precut the remaining shingles up the roof. This idea works in a perfect world, but the wall above the roof may bow in or angle out an inch or more. The neatest and most effective way to butt the shingles in is to cut the shingle for each course as you go up the roof. It takes more time to cut the shingles one course at a time, but then there isn't a widening gap of step flashing showing off the cut ends of the shingles on up the roof.

TYING INTO A WALL ON AN OVERLAY

It is fairly common practice when overlaying a roof to butt the new shingles against a wall without weaving new step flashing in with the new shingles. *Consider leaving out the new step flashing only if you have never experienced leaking beneath that section of wall.* An intermittent leak, such as one that occurs only during storms with high winds, indicates a damaged, missing, or improperly overlapping piece of step flashing. The way to correct that leak problem is to install new step flashing as you overlay the roof.

Nail: If you have experienced leaking along a wall, some roofers will try to get by using caulk or mastic to seal the new overlaid shingles to the wall. This is a lousy repair: the leak will return with amazing speed.

In preparation for overlaying the roof, you have cut the third course and exposed the bottom of an original piece of step flashing. Loosen and remove the nail in this exposed piece of step. Set two (or three, if necessary) pieces of new step flashing in place over the new starter shingle you are ready to lay. The top piece of new step should slide 2 inches under the piece of old step flashing. Make sure the laps on the new pieces of step are all at least 2-inch downhill laps, and that the new step extends to the bottom edge of the starter shingle. Nail the pieces of new step in place. The nail in the old piece of step should go back in the original nail hole. What this additional step flashing does is kick any water running along this wall up on top of the new shingle at the fascia board. Now, you no longer depend on the old bottom shingle to turn the moisture for another twenty to twenty-five years.

Nail: It isn't absolutely necessary to kick the step up on top of the new starter course like this. Most roofers don't do it. They just butt the shingles to the

wall all the way up with no new step flashing anywhere.

If you've never had a leak problem, you don't want to start one. As you continue to lay new shingles on top of the old roof, remember the original step flashing comes out 3 inches under the old shingles. Make sure the last nail toward the wall doesn't go into the original step flashing. This lets the original step flashing continue to do its job undisturbed. Don't worry about not nailing the new shingles the last 3 inches against the wall. The wall rising above this lower roof is protecting that end of the new shingles from wind uplift.

If you make a mistake and nail through the original step flashing, try not to worry about it. Chances are, it won't leak. It's just a better idea not to nail through it and ask for trouble.

INSUFFICIENT ROOM FOR VERTICAL LEG OF NEW STEP FLASHING

You have experienced a leak in the past, so you will install new step flashing. You get partway up the roof with the new step flashing and discover you can't slide the vertical leg of the step flashing under the wood, aluminum, or vinyl siding of the wall. There are a couple of tricks to try. If the siding is aluminum or vinyl, use one hand to pull the siding slightly toward you while you slip the pieces of step into place behind it with your other hand. If that doesn't work, you may be able to tell that the piece of step is being stopped by a nail. (I think siding contractors drive a few nails down low so new step flashing can't slide into place. They do it just to be ornery.) Rock the piece of step from one side to the other until you are sure where the siding nail is located. Take the snips and cut a small notch out of the top of the 2-inch leg of the step flashing, then slide the notched 2-inch leg around the shaft of the nail. You're not likely to have a flood of water flowing along this wall, so the notch won't hurt anything as long as you don't completely do away with the leg that goes up the wall. Use your good judgment here. If you notch the step flashing all the way down to the "L" bend, you've broken the continuity of the step flashing, and it will probably leak.

TOO LITTLE GAP BETWEEN SIDING AND SHINGLES

A good siding man is going to keep his nails out of the way of the step flashing. He is also going to cut his siding up high enough that a roofer will have an easy time installing step flashing on an overlay. However, there could be a sequel to this book showing "Nails" for siding contractors. Too many siding contractors cut their siding so that it is almost snug down on the shingles themselves. This makes it impossible to slide new step flashing in place when you overlay the roof.

If you have a leak along a wall and the siding is down on the shingles, you have two options.

The first is to cut the old shingles in a line just off the end of the old step flashing. Be careful not to cut the "paper" or roofing felt beneath the shingles and step flashing. You can continue to get some use out of this old felt. Remove the cut pieces of shingle from the old step flashing, starting at the ridge of the roof. Carefully pull the nails and leave the old step flashing in place. After the cut ends of the old shingles are removed from the step, shingle up the roof from the starter course, weaving the new shingles into place using the old step. Reuse the original nail holes in this old step flashing. No need to ask for trouble punching extra holes.

When you cut away the old shingles along the wall, you may discover that this gives you enough room to remove the old step flashing and install the new pieces. This is preferable to reusing the old pieces.

Either one of these methods causes the new shingles to dip along the wall. Your new roof won't look as good as it would have if you hadn't been forced to cut and remove the old shingles along the wall.

If the old step flashing is galvanized, it is probably already rusting. Don't trust it for another twenty years; it won't make it.

If the old step flashing is copper, and you are going back with copper on the valleys and other trim, reuse the copper step flashing. If the copper is in good shape, it should easily last the life of the new roof. One factor in the decision to reuse the copper is that the price of each piece of copper step is ten to fifteen times the price of the aluminum.

The carport has a lean-to roof.

Turn the shingle over and cut along the chalk line. Turn the shingle right side up and nail it in place.

Twenty-four-inch painted aluminum has been bent in a metal brake.

I notched out for the window sill, then I cut the rest of the back down so the aluminum flashing will slip behind the siding.

I bedded the lower edge of the flashing in a heavy bead of mastic and centered each nail over the tab below. I replaced the siding, then put a dab of silicone caulk over each nailhead.

One way to tie in the top piece of step flashing. Note the copper piece that will go in front of the V-notch.

15.
Brick Chimneys and Skylights

You are probably worried about flashing your brick chimney. Chimneys come on like an "800-pound gorilla" and are responsible for a large number of do-it-yourself roofing projects which were never started. The method we will use to flash and tie in a chimney is very straightforward: you will be pleased with the outstanding (no-leak) results. You will be shown the way up and around the chimney, and my figures will put you on the roof doing each step. Someone wise once said, "Worry is sadistic entertainment." He was right. We've come this far together, so relax and let's enjoy ourselves as we tie-in a chimney.

There are several steps to flashing a chimney. It takes time to trim and fit the various components. It's not complicated, it just takes time. You can easily spend four to six hours on your chimney, so get rid of any idea that you must complete it in two hours. The test of your work is not how fast you do it, but whether it leaks and whether it lasts.

It is best to have the upper and lower bent metal flashing already bent to the angles you need before you tear the roof off. I used 24-inch (width) coil, 0.027 (27 gauge) aluminum for these pieces. Let the bottom and top metal extend 6 inches beyond both sides of the chimney. If the chimney is 2 feet wide, make the metal 2' + 6" + 6" = 3' long. If you want the bottom flashing to rise up the brick 6 inches under

the skirt flashing, mark the coil and bend the metal at an angle slightly less than the one between the sheathing and the lower face of the chimney. Remember, you want the metal to tend to spring toward, not away from, the brick face and sheathing.

The metal above the chimney will be bent at greater than a right angle (90°). Don't overbend this metal either. Snow and ice tend to accumulate above the chimney, so make sure you have a substantial leg (6 inches plus) protecting the back face of the brick chimney.

How else do you make the job easier? Let's say you are doing a tear-off. The far side of a chimney is even with the left rake, but the remaining three sides are out in the shingles. Go ahead and tear off the right side of the front roof, but *stop* the tear-off on a rough vertical line 3 feet before you get to the chimney. Just leave the left end of the old roof intact from the ridge all the way down to the fascia.

Felt in, mark, and shingle the right portion of the roof. Roll runs of felt out 2 feet over the edge of the old roof and weight the ends down with bundles. Stop the new shingles about 1 foot from the edge of the old shingles and lay the ends loose. If the forecast at the end of the day has suddenly changed to rain for that night, you can run a heavy bead of mastic down the edge of the felt, overlapping the old shingles. This will help prevent rain from blowing in under the side. Weight down the edge over the mastic bead. The next day when you are ready to concentrate on the chimney, cut off the excess felt overlapping the old shingles. Just make sure you leave at least 6 inches of felt sticking out from under the new shingles. Complete the tear-off to the left rake. Lap the ends of the roll of felt the required 6 inches over the felt you previously laid. Complete the runs of felt to the left rake or the chimney.

If the chimney is at the rake, you may only need to flash the bottom, one side, and the top. If three sides of the chimney are outside the wall, you may have to flash only one side of the chimney. For our example, we will complete a chimney that is midway across and midway up a roof. You will know how to flash it all the way around. Then you can eliminate the parts that don't apply to the chimney on your roof.

The Building Code calls for No. 30 felt below and above the chimney. I prefer the additional protection of modified bitumen (or 90 lb. rolled roofing) all the way around the chimney. Starting and stopping No. 15 felt to insert short sections of No. 30 felt (or in our case, modified bitumen) at the chimney is a lot of extra trouble. It is easiest to do a normal felting in and then come over the No. 15 felt with the modified bitumen. So when a run of felt is interrupted by the base of the chimney, cut out a large rectangular "notch" for the chimney, so the cut sides and bottom of the run of felt will lie flat on the sheathing. When you are past the chimney, continue the run on across the roof. Trim the next run or runs of felt into the sides of the chimney. Lay the run of felt across the top of the chimney too, notching out a rectangular pattern for the upper portion of the chimney.

Hint: If a sudden squall should come up at this point, seal the edges of the felt to the chimney with mastic. Weight the felt down, and it will do a fairly decent job of keeping the rain out.

Roll the modified bitumen out below the chimney and cut it so that at least 1 foot extends beyond each side of the chimney. Slide the modified bitumen up

the roof until approximately 6 inches of it runs up the face of the brick chimney. Use the straight blade to cut straight down the modified bitumen at each corner of the chimney. You want the 1 foot on either side of the chimney to lie flat on the sheathing and the portion you cut free to lie up the face of the brick. Don't make the cuts exactly in line with the sides of the chimney. Shift the location of the cuts 1/4 inch in toward the center of the chimney. This will cause the 1-foot legs up each side of the chimney to fit snug against the base as shown in Figure 9. Nail the bitumen in place, keeping the nails well away from the chimney itself.

(You can cut off the 6-inch portion that climbs up the brick face of the chimney. Cutting off the 6 inches up the face makes the finished metal look a little neater, but it also eliminates a protective feature across the base of the chimney.)

Trim two 1-foot lengths of bitumen and lay a length into each side of the chimney. The 1-foot lengths overlap the 1-foot legs of the first piece of bitumen

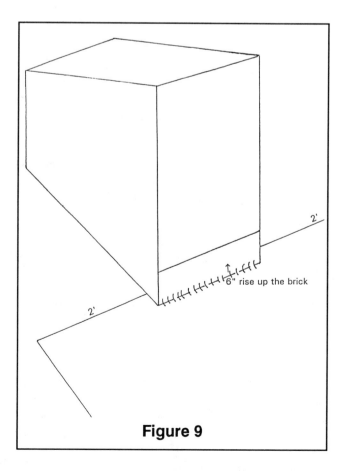

Figure 9

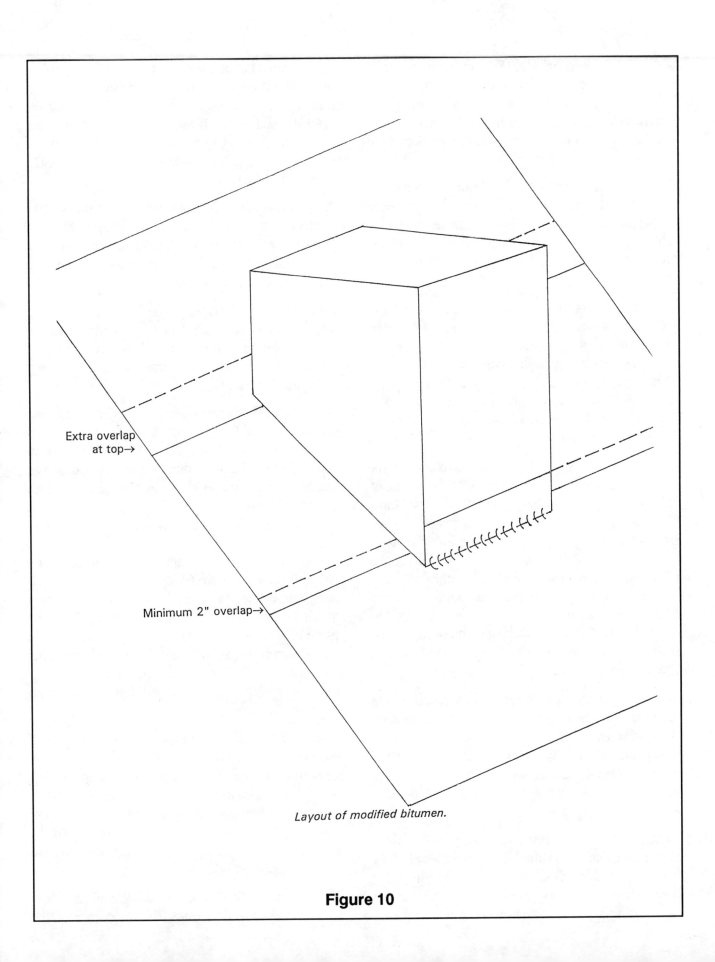

Extra overlap at top→

Minimum 2" overlap→

Layout of modified bitumen.

Figure 10

with a 2-inch vertical lap. If the chimney requires additional 1-foot lengths of bitumen up the side, go ahead and lay them. If the sides of the chimney are relatively short, allow more than a 2-inch downhill overlap. It's just that much added protection. Nail all the side pieces in place.

When you reach the top of the chimney, cut a piece of modified bitumen the same way you did the piece across the bottom of the chimney, and let the 1-foot wide legs drop down over the 1-foot side pieces. Nail down the modified bitumen as in Figure 10.

Lay the courses of shingles on up to the chimney. If you used a 24-inch coil stock to shape the metal and you bent 6 inches to climb the lower face of the brick chimney, you are going to have 18 inches of metal showing over the shingles beneath the chimney. This length of lower leg provides maximum protection against rain blowing back in under the lower bent flashing. However, 18 inches is a lot of metal to be showing on a highly visible roof.

You will probably decide to trim this lower leg, for aesthetic reasons. Let's say the top of the course of shingles just butted into the lower face of the chimney. That puts the tops of the keys down 7 inches from the chimney. You want the bottom edge of the metal to come out onto the finished surface of the tabs 2 inches. A lower leg with 9 inches over the shingles will look better than one with 18 inches. Mark the metal 9 inches back from the lower milled edge. Pop a chalk line at the 9-inch marks and cut 9 inches off the lower leg. (This is a finished cut, and it will show.)

I prefer to have the milled edges of the coil showing on all finished edges. I set up the edges I cut with shears so they were under skirt flashing or covered with caulk. My cuts were fine, but the mill finished edges were cleaner. However, this isn't always possible to do when you form or bend the metal in advance.

Center the bent metal flashing on the bottom face of the chimney and mark the 6-inch upper leg slightly inside the corners of the chimney. Cut the 6-inch leg perpendicular from the marks to the bend. Set the metal back in position and push the 6" x 10" side

legs you just cut at each end, flush with the surface of the modified bitumen. (As with the modified bitumen, you want the legs to fit snugly against the brick sides. It's fine if the legs tend to "climb" the brick a little bit.) Lightly hammer the bends down straight on the 10-inch side legs so they will lie flat on the modified bitumen.

As you keep the 6-inch long upper leg that climbs the face of the brick chimney pulled up tight to the chimney, nail the 10-inch leg down. An aluminum nail should be in each 10-inch side leg at the top corner farthest from the chimney.

Run a bead of mastic under the lower bent flashing edge which is out over the exposed portion of the shingles. Nail this lower edge down, centering the nails over the tabs of the course it covers, as shown in Figure 11.

Bring the next course of shingles in. Cut the shingle into the side of the chimney the same way we did when tying into a wall. The bottom piece of step flashing should extend out over the bottom leg of the lower bent metal flashing by 2 inches. This bottom piece will also have to fit up under the skirt flashing you will install around the chimney. You trim this bottom piece of step the same way you did the step in Figure 8 in Chapter 14. Figure 12 in this chapter shows the first side course tied into the chimney. Remember to add a second piece of step directly on top of the shingle and above the trimmed piece of step. Keep the bottom edge of the second piece of step just above the self-sealing strip. This will give you the proper downhill overlap for the remainder of the step flashing in the courses on up the roof.

Figure 13 shows the courses tying into the side of the chimney. Don't nail into the shingles if the keyway is at the step. Let the nail in the step flashing hold the shingles down. Notch the 2-inch leg of the last piece of step flashing so that it follows the corner of the brick. Leave a small section (3/8 inch) of the 2-inch vertical leg in place at the bottom of the area you notched out.

If the top of the final course of shingle up the side is at a higher elevation than the top corner of the

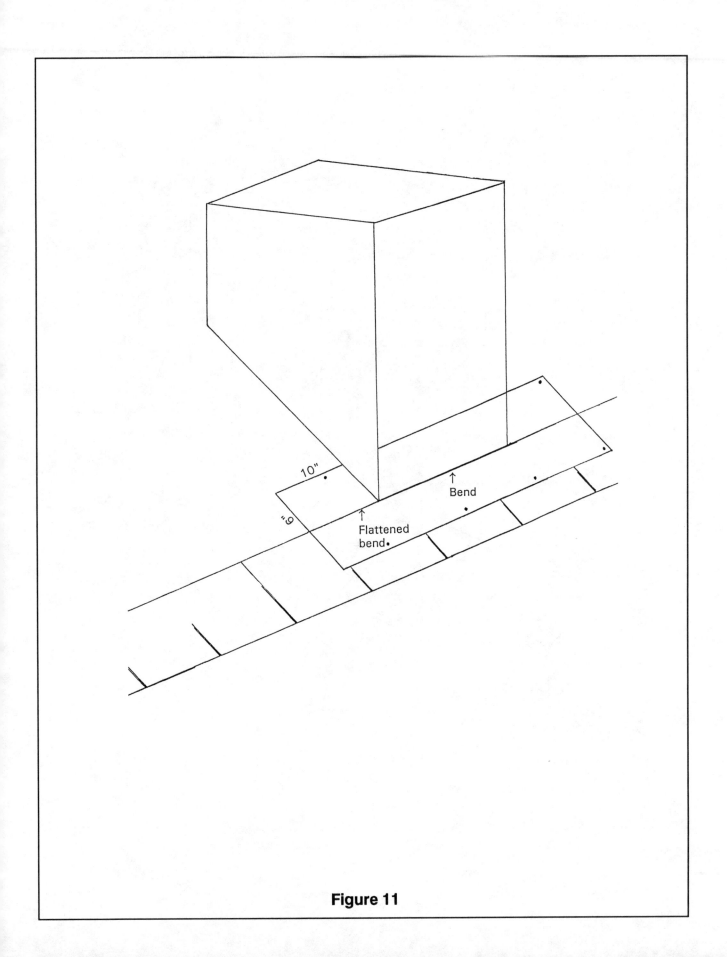

Figure 11

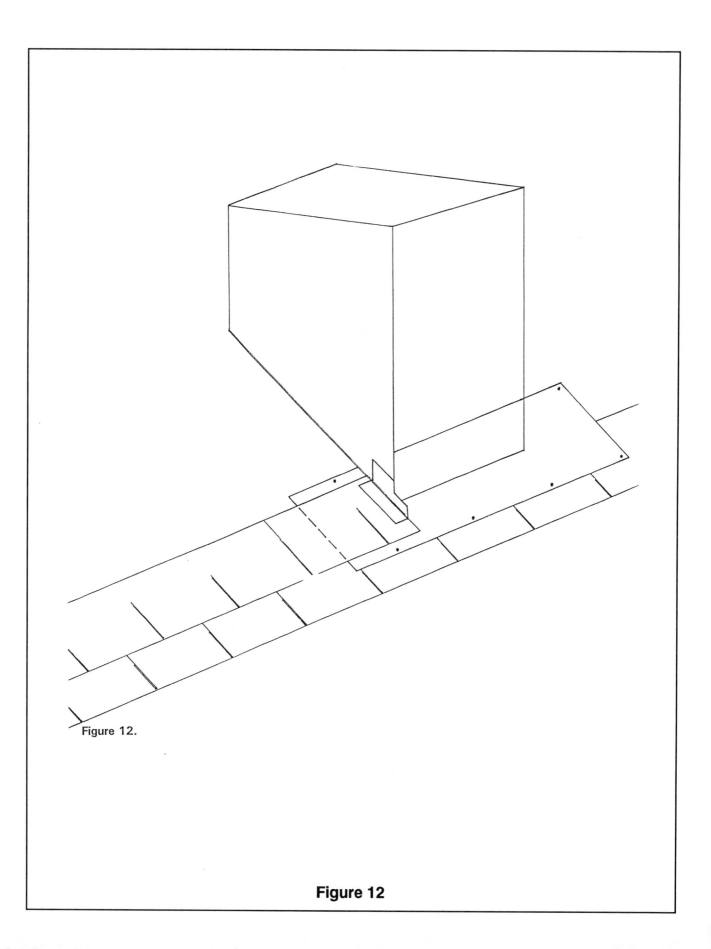

Figure 12.

Figure 12

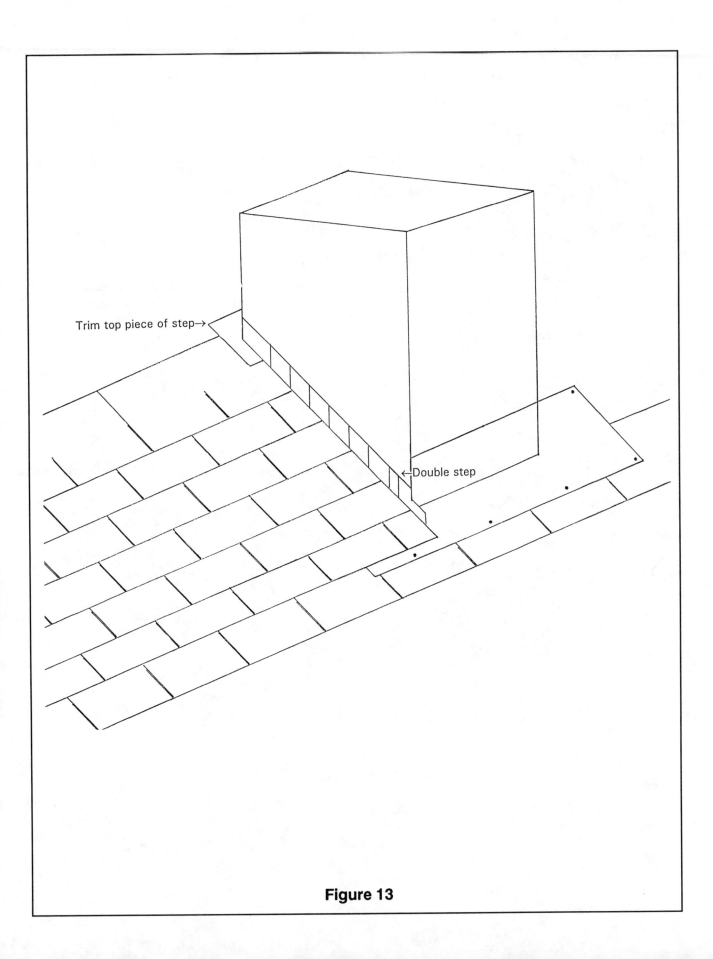

Trim top piece of step→

←Double step

Figure 13

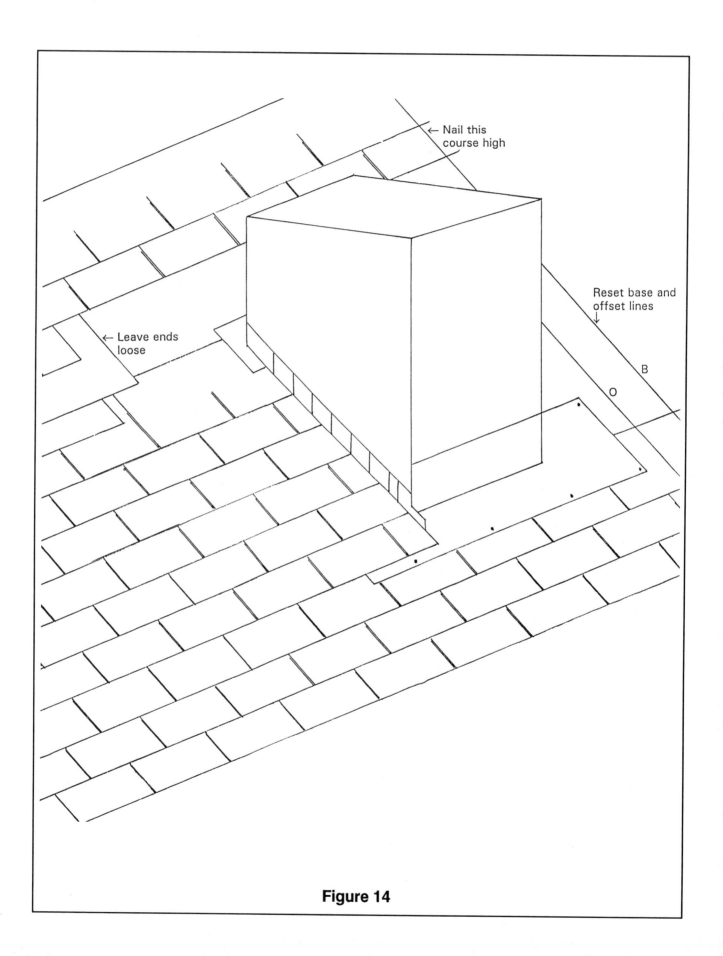

Figure 14

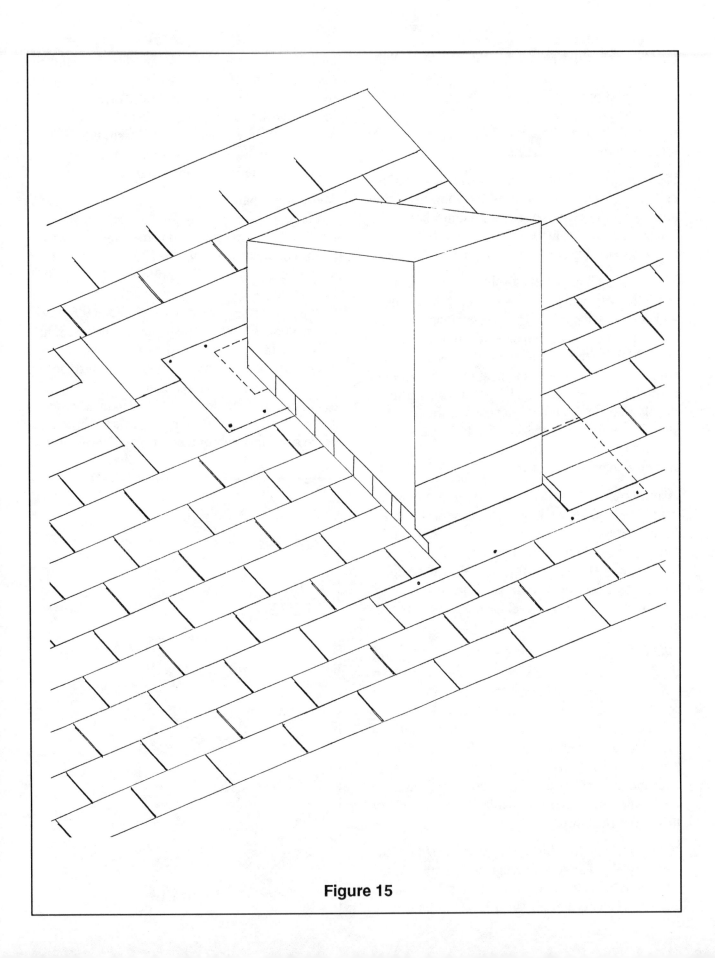

Figure 15

chimney, trim out as much of the bottom of the shingle as you have to and carry the few inches of the top part of the shingle over the modified bitumen all the way across the top of the chimney. This will raise the upper bent metal flashing and keep it level with the rest of the roof.

Reset the base and offset lines on the far side of the chimney (Figure 14). This is the same as going around a power ventilator as we did in Chapter 12. Tie in the shingles on the far side of the chimney all the way to the top corner as you did on the left side.

Place the upper bent metal flashing above the chimney with its 6-inch leg up the chimney. Mark the 6-inch leg 1/4 inch *outside* the corners of the chimney. Cut from the marks perpendicularly down to the bend. Making the cut for the leg down the chimney slightly long will cause the bent metal flashing to carry the runoff water slightly beyond the corner of the chimney and out onto the base of the trimmed step pieces at the top of each side. Push the 10-inch legs down over the base of the top pieces of step flashing on each side of the chimney.

Hammer the bends on the 10-inch legs down flat. Nail each 10-inch leg down flush, driving the nail in the corner farthest from the chimney. Caulk under the top edge of the upper bent metal flashing and bed it down to the felt. Nail the top edge of the upper bent metal flashing down to the felt and sheathing, keeping the nails as far from the chimney as practical. (See Figure 15.)

You will notice that by carrying the 3-inch base of the first piece of trimmed step flashing a few inches out on top of the bent metal flashing at the bottom of the chimney, we have carried most of the water that will run along the side of the chimney out past the lower corner of the chimney. By extending the 3-inch leg of the top piece of step flashing under the bent metal flashing above the chimney, we have intercepted the water that will roll off the ends of this upper flashing.

Next, we shape the skirt flashing on the side of the chimney. We want it to overlap the front face of the bent metal flashing that goes up the face of the chimney. Cut off the coil a length of aluminum long

enough to rest on the roof and bend around the corners of the chimney, with a surplus 3 inches overlapping the upper and lower faces. Bend the lower overlap to the shape of the corner of the chimney. You will have to cut up slightly into the coil to make the 3-inch overlap in front rise above the bend in the lower bent metal flashing.

Sight down behind the metal and see where the mortar joints are. You are going to trim the metal so it follows the horizontal and vertical mortar joints in the brick. Your metal will have a stair-step pattern along the top when you are finished. To make it fit the bottom, you will have to cut up into the coil slightly from the corner bend. To make it fit at the top, you will have to cut from the bend slightly downward toward the bottom of the coil. You will mark the beginning and end of the steps by dimpling the center of the horizontal and vertical joints with a nail and marking the lines between these dimples with a straightedge. (Figure 16 shows a profile view of the skirt flashing for the left side of the chimney.

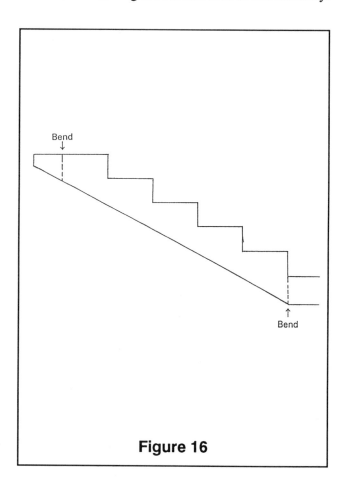

Bend

Bend

Figure 16

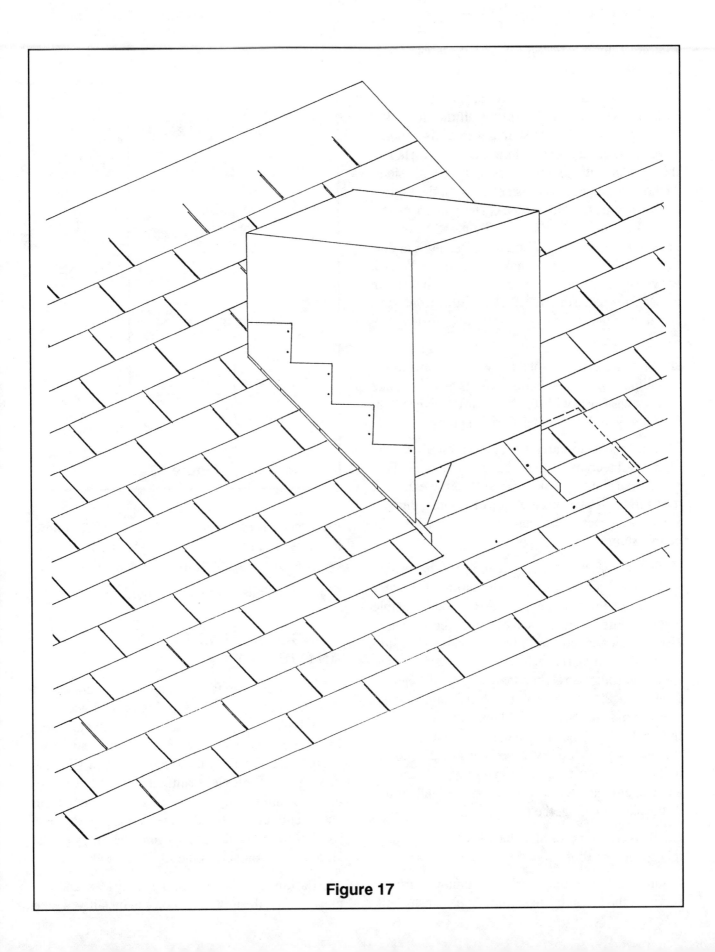

Figure 17

Figure 17 shows the chimney with the left and right skirt flashing installed.)

Once the left and right skirt flashing have been installed, you can shape and install the front skirt flashing. Cut the length so that it extends 3 inches beyond the corners of the chimney. Cut it to a height that ties in with the skirt flashing on both sides. Make sure the milled finish edge is on the bottom where it will remain visible. (The pattern of this front skirt flashing is shown in Figure 18.) You will have to cut very narrow notches at the bottom corners of this front piece to make room for the ³/₈-inch leg you left in place on the first piece of step flashing at each corner. (Figure 19 shows the front skirt flashing in place.)

The skirt flashing for the back of the chimney is shown in Figure 20. Note that you will have to cut a "saddle" up into the bottom of this piece to make it fit down flush with the bend in the upper bent flashing. Install the upper skirt flashing.

You are ready to carry the courses on above the chimney. I recommend you leave 2 to 3 inches of the roof leg of the upper bent metal flashing exposed. The wind will have an easier job of blowing dirt and leaves off the smooth metal than it will blowing the coarse shingles clean.

With the overlapping metal flashing, you have done an excellent job of mechanically turning the water away from the chimney. There are four possible trouble spots — one at each corner of the chimney. Use your finger and carefully knead caulk back into the small openings at each corner. (Be careful not to rake your finger across the sharp edge of the metal.) Now caulk the entire top of the skirt flashing all the way around the chimney. Pull the metal away from the brick and let the bead of caulk get down behind the metal too. Caulk over the heads of the mortar nails so they won't rust. When you have caulked the small openings at the base of each corner and the top of the skirt flashing, the chimney can't leak.

Caulk over any exposed nailheads when you finish this section of roof.

Figure 21 shows the finished chimney ready to serve for the life of the new roof. Apply fresh caulk

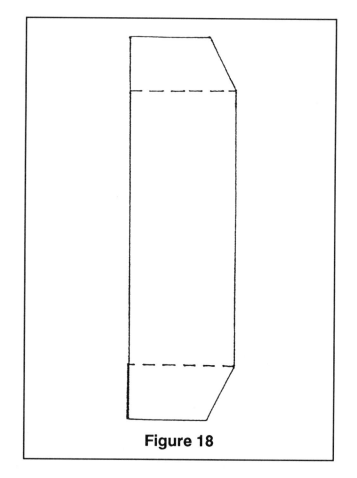

Figure 18

when you do the overlay two decades from now, and this flashing will last beyond the twenty- to twenty-five-year life of the overlay too.

Congratulations! I knew you could do it.

TYING IN A CHIMNEY ON AN OVERLAY

When you get up to the chimney with the courses of shingle, gently raise the roof leg of the lower bent metal flashing. You may have to pry up the whole first course beneath the roof leg to keep from bending the metal. Pull the nails out and slide the new course of shingles in underneath the lower bent metal flashing. This keeps all runoff water from above and around the chimney on top of the new roof. Now butt the courses into the sides of the chimney and on across the top.

The flashing around the chimney may leak, and you may not find the problem. You can install new step

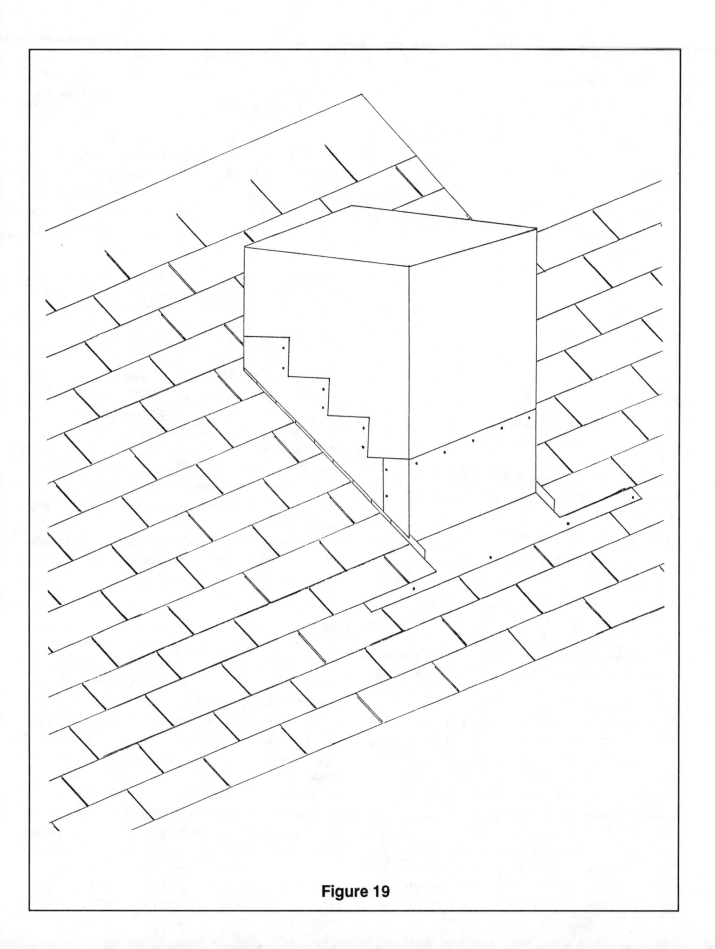

Figure 19

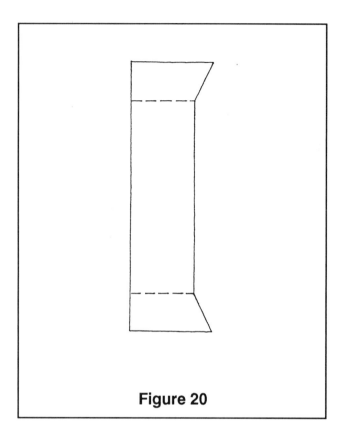

Figure 20

flashing with the new shingles. Lap the first pieces of step flashing out over the roof leg of the lower bent flashing and tuck the final piece of step up under the roof leg of the upper bent metal flashing.

If you don't have room to slide the 2-inch leg of the step up under the lower edge of the skirt flashing, you can completely redo the flashing system. It's neater to tear the whole skirt flashing off, but you can completely redo the metalwork right over the original metal. The old flashing and old shingles underneath will provide some minimal additional protection.

Many, many roofers will overlay a roof around a chimney and won't redo the flashing. It's okay if the flashing is in good shape, if he chips off and redoes the caulking, and slides and caulks the new shingles in under the lower bent metal flashing.

Nail: Unfortunately, some roofers don't raise the roof leg of the lower bent metal flashing and lay the new shingles in under that leg. They just cover the roof leg with the new shingles and cut the shingles around the chimney. When they do that, the com-

plete flashing system is covered. Water that penetrates to the old original step flashing runs down beside the chimney and continues down the roof *underneath* the new shingles.

Your roofer in effect has punched the old shingles down the roof from the chimney full of nail or staples from the new shingles and expected the old roof to remain in service another twenty to thirty-five years. The homeowner will soon experience mildew on the ceiling at or near the fireplace, and the smell of rot will become noticeable.

Incidentally, not tying in under the flashing saves the roofer an absolute maximum of ten minutes.

METAL-FRAMED SKYLIGHTS

Skylights raised above the shingles of the roof on a 2 x 4 box with a sealed metal frame are flashed the same way as a chimney. The metal framing sealed to the glass of the skylight drapes down over the 2 x 4's and performs the same function as the skirt flashing on a chimney. The upper and lower bent metal flashing are bent and trimmed to fit the same way you did it on the chimney, and the step flashing along the side ties in the same way. The only difference is, instead of making your own skirt flashing, you trim the legs up the side of the 2 x 4 box so they will fit under the "skirt" of the metal framing of the skylight.

Framed skylights are much more costly than the bubble skylights with the 2½- to 3-inch lip molded around the outside of the bubble. These bubbles are tied in with mastic and more mastic between sheathing and felt, felt and molded edge, molded edge and shingles, and finally mastic between the edge of the shingles and the bubble of the skylight. The bubble is guaranteed against manufacturing defects for so many weeks or months. The only guarantee I can give you on the bubble type is that it's designed to leak and that is exactly what it's going to do. There is a significant price difference between the better quality raised metal-framed skylight and the cheap bubble type. If you install a bubble skylight and it never leaks, your kind of luck belongs in Las Vegas.

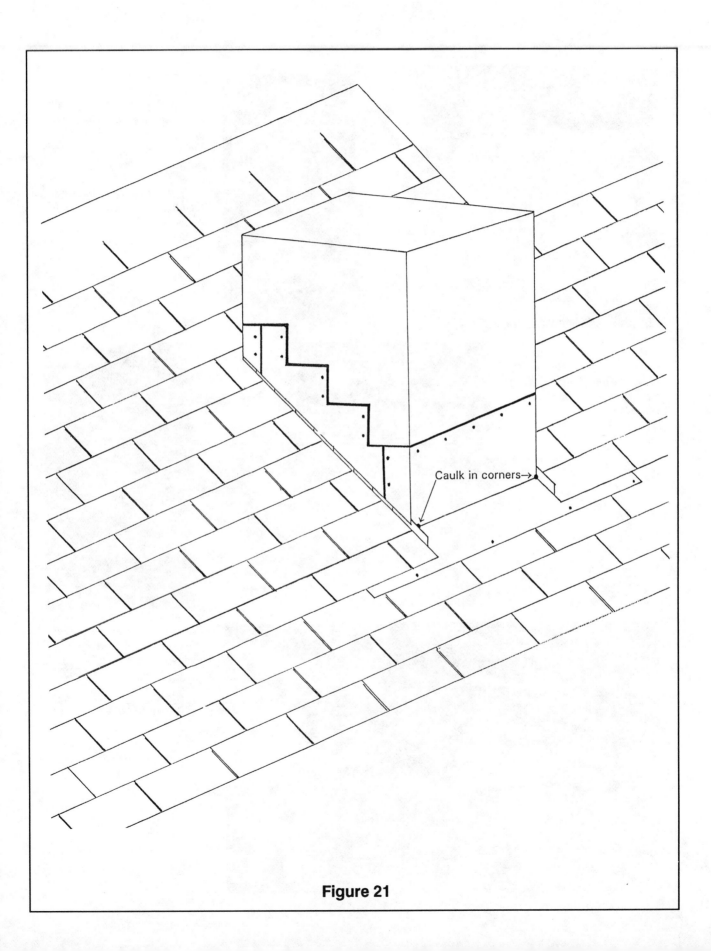

Caulk in corners→

Figure 21

This brick chimney is partially outside the exterior wall and also penetrates the roof. Lay the shingle right up to the chimney, and run Number 30 felt to the chimney. Cut the lower flashing and bend the cut leg down flush with the roof. Bed the bottom edge of the lower flashing in mastic and nail it in place.

I fabricated this first extra-long piece of step flashing on the brake. I wanted the white step flashing coming out over the white lower flashing.

Notch the lower part of the shingle for the chimney.

Cut the top piece of step over the chimney.

Place the upper flashing, cut the leg, and bend it down the roof.

Bring the next shingle over the upper flashing.

Hold your metal up against the chimney and sight behind the metal. Dimple the metal at the various breaks in the mortar joints. Lay the metal flat and scribe (scratch) the metal with an awl or a nail along a straight edge connecting the dimples. You will end up with the exaggerated stair-step pattern I used on this piece of skirt flashing up the side of the chimney.

Mold the bottom of the skirt flashing around the chimney corner, pull it away, and cut the slit for the small vertical leg of the step flashing.

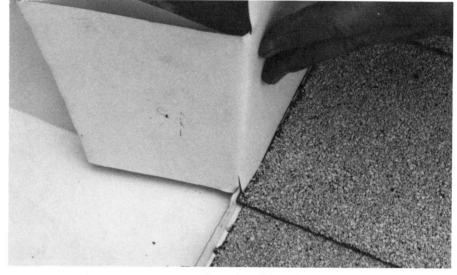

Puncture the metal with a nail to keep your drill bit from "dancing."

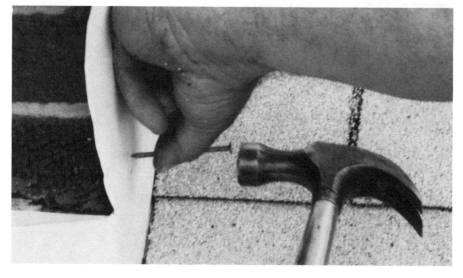

Pre-drill the holes for your mortar nails. Use a mortar bit slightly smaller than the diameter of the shaft of the mortar nails.

Install the front piece of skirt flashing.

Caulk the upper edges and vertical joints with silicone. Silicone is sticky and will start setting up in less than ten minutes. Smooth it with the flat surface of a piece of step. Caulk over all the nail heads. Do not caulk across the bottom of your skirt flashing or lower flashing. Doing so will form a dam. Knead a dab of caulk in the corners. Otherwise, there will be a small opening where the flashings overlap.

The Code doesn't require that a 7/12 roof have felt on it. I wouldn't do a roof without felt. I rolled the felt and modified bitumen both up the chimney. I carried the modified bitumen out several inches to each side. A steep roof doesn't warrant quite as much extra protection around the chimney as a lower slope roof. I let the lower edge of the modified bitumen overlap the course of shingles.

The next course of shingles goes right over the bitumen.

The lower copper flashing is in place, with the legs flattened up the roof.

Tying in from the loose ends to the copper step flashing of the chimney (copper nails in the copper step).

Going up!

The other side. I delayed completion of the chimney for several days while I worked on less time-consuming items. Note the mastic sealing the top edge of the modified bitumen.

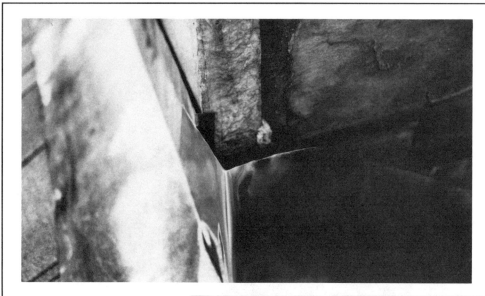

Skirt flashing overlapping from the side to the front.

I left the side skirt flashing straight on top rather than stair-stepping it to try to match the erratic mortar joints in the stone.

Other side and lower flashing.

I fed black silicone caulk in behind the upper edge of the skirt flashing. Finishing beads of caulk went on as soon as I rough-caulked each side. (Remember the quick curing time.)

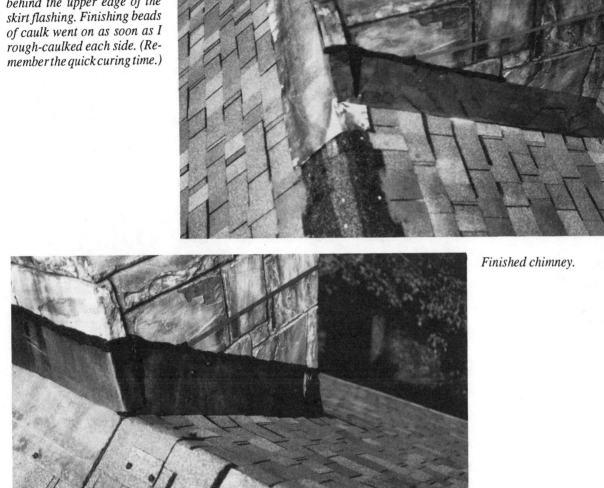

Finished chimney.

Nail: *The ends were laid loose around this chimney. The contractor never came back and nailed down, so over time the shingles have sagged down the roof.*

I installed three 21" x 27" skylights in the family room and garage wing. The rafters are 2 x 6's on 16-inch centers. I prefabricated six replacement jack rafters with offset spacers on each end. Each replacement is an 8-foot 2 x 6 with a 2 x 6 and a 1 x 6 spacer nailed on each end with 16d nails. I then predrilled four bolt holes on each, being careful to zigzag the pattern of the holes along the 8-foot 2 x 6 so it wouldn't split when I tightened the bolts.

I popped a chalk line along the face of the replacement rafter, set my circular saw for a ³/₄-inch depth, and cut away the old plank sheathing.

I used double 2 x 6 headers (top and bottom) to tie the rafters together across the opening. I trimmed some of the old 1 x 8 sheathing and nailed it on top of the headers to restore the top and bottom of the opening to the original level of the sheathing.

I installed the brackets to the skylights before I took them up on the roof. A nail through a screw hole in a bracket on each side of the skylight kept it from sliding off the roof. Drill all the screw holes into the sheathing and screw the skylights in place.

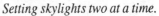
Setting skylights two at a time.

Restore the felt up from the bottom. Curl it up the bottom wood framing of the skylight and let the leg drop up the roof. Paper the sides.

Paper the top. Tuck the paper under the next seam of the old paper up the roof. Don't try to use modified bitumen on these skylights. The tolerances on the manufacturer's flashing kit are way too close for material that thick.

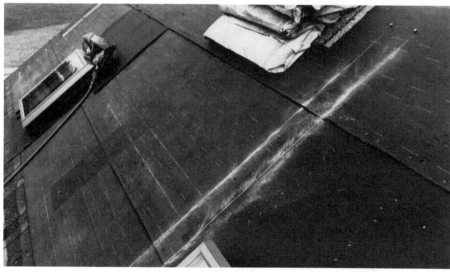

I had to restore my original base line and offset line because thunderstorms washed them away before I cut the holes for the skylights. Tip: I popped the horizontal course lines before I cut the holes for the skylights. Popping the lines first saved establishing the short sections up the sides of the skylights.

The lower and upper flashing in the manufacturer's kit was pre-formed. I removed the screws from the sides of two lower brackets, raised the skylight unit up, and slid it back down into place over the lower flashing. I refastened the lower brackets, then removed the screws from the sides of the upper bracket. I raised the top of the unit and lowered it over the pre-formed upper flashing. This was all necessary to get the vertical portion of the lower and upper flashing under the continuous rubber skirt that runs completely around the lower side of the skylight unit.

Mastic under the top edge of the upper flashing. Then nail across the top edge of the upper flashing.

Cut the shingles that go on top to fit, then mastic the upper surface of the flashing.

Always pop new verticals once you are past the obstruction. I stopped the two courses short of the rake and drove a nail directly above the end of each shingle. I held the chalk line down on the keyways down the roof and popped the lines. The shingles on up the roof above the skylights are laid from my original base and offset lines. Popping these new verticals beyond the obstruction keeps the keyways straight all the way up the roof.

When I did the other side of the skylight, I ran a heavy bead of mastic beneath the lower edge of the bottom flashing and nailed it in place (aluminum nails). I put the first piece of step flashing in under the rubber skirt and slid it down as far as it would go. (Don't nail it.)

Next course — don't nail it.

Next piece of step — the nail through the bottom of this piece of step caught the first step and the end of the shingle.

Next course, and then add the next piece of step.

At the top, slide the step in under the leg of the upper flashing and nail it. Slide the next course of shingles in under the leg of the upper flashing.

It is now a tight squeeze at the top, so trim the top ½ inch off the final piece of step. You will have to remove completely the upper 2 inches to the vertical leg of this piece of step. I left the part to be removed hanging by a thread so you could see it.

I slid the final piece of step in place and nailed it.

16.
Valleys

You should install a metal valley on the roof. You can go to a "Double Weave" or a "California Cut" valley if you want to, but a metal valley is far nicer looking and it's definitely more durable.

No matter what surface finish you use for a valley, the underlayment is done the same way. The Building Code calls for *No. 30 felt* to be run down the flow line of the valley and the standard No. 15 felt laid into it from the sides. A better alternative to No. 30 felt is modified bitumen or rolled roofing.

Modified bitumen comes in a roll and is heavier and more expensive than rolled roofing. As you saw when we flashed the chimney, I prefer to use the modified bitumen. However, we will use rolled roofing to line our valleys in this chapter.

Rolled roofing is either asphalt and fiberglass or a rubberized compound. It has a grit surface like shingles. Rolled roofing is designed as a finished roofing surface itself. It is aproximately 40 inches wide and rolls out like the roofing felt. Rolled roofing is thick and tough. It won't tear, it won't get a hole punched in it, and it won't fail in the valley. Rolled roofing will seal around the shafts of nails that are driven through it. You will work harder because of the weight of the rolled roofing, but the valleys will be trouble-free. Rolled roofing is fine in place of modified bitumen, and both are much better than using No. 30 felt down the valley.

One bad thing about valleys is that, theoretically, you need to tear off and rebuild both roofs that meet

at a valley at the same time. Problems occur when you are working alone and both roofs are sizeable. It is impossible to tear off both roofs and lay them and the valley back up in one day. It is true you can do a good job of felting or papering the roofs in to "carry" them overnight. However, avoid planning on doing this routinely. If you tear off both roofs and the valley, it means you will have to leave a roof open (with bare sheathing) or protected only by the felt overnight.

You can use the trick discussed in the last chapter. Tear off part of the two roofs that meet at a valley. Leave part of each old roof in place from the bottom of the valley straight up both roofs to the ridge. Lap the new felt over the old roof 2 feet as we did in Chapter 15 with our chimney. Then the next day (or the day after that), you can tear off and replace the valley, concentrating just on the valley itself and the shingles immediately adjacent to it.

Let's say you have "certified" good weather for a long weekend, and you have torn off both roofs and the valley. The first thing you do is unroll the rolled roofing *up* the valley. Let the loose end of the roll overlap the fascia board by at least 1 inch at each corner of the roll. (The center of the roll will be several inches over the end of the flow line of the valley.) Keep the roll centered over the flow line as you unroll it up the roof. If you try to run the roll down the valley, the weight of the roll will pull you down the roof. The roll is going to get away from you and go completely off the roof, possibly taking you with it. So let the loose end drop over the fascia, then keep the roll above you and push it up the roof. Straddle the rolled roofing. Remember, it isn't fastened (plus some modified bitumen has a very slick surface). Once you are over the ridge of the wing roof, brace the remainder of the roll against the valley coming to the ridge from the other side.

You may have to shift the rolled roofing to center it up the flow line or to smooth out ripples. Push the rolled roofing down snugly into the flow line. Nail both edges of the rolled roofing to lock it in place.

Cut the bottom end of the rolled roofing to the shape of the intersecting fascia boards. Leave a 1-inch overhang over the fascia boards of both roofs leading to the valley. The rolled roofing is the backup system to catch any water that may escape the valley. You want it to carry the water beyond the face of the fascia board.

Now go back up to the cap at the top of the valley. Drape the rolled roofing over the ridge and cut it parallel to and 6 inches beyond the ridge. When you have cut to a point opposite the top of the valley, cut the rolled roofing straight up the main roof. Make a cut back down into the roll from the cut edge to within 1/2 inch of the top of the ridge. (See Figure 22.) This cut down to the top of the ridge will let the 6-inch overlap lie flat in the far valley. Leave this top end beyond the ridge loose (don't nail it down) for now. Pull the remainder of the roll free and brace it out of your way.

(When you have torn off the far side of the wing and the far section of the main roof, you will bring the rolled roofing up the far valley. Cut the top of the roll from the far valley the same way, and tuck it under the rolled roofing for this first side of the valley. Nail the roofing from the far valley in place over the ridge. Nail the rolled roofing you left loose at the top of this valley over that of the far valley and nail in place. This keeps the exposed edge of the overlap under the less visible metal valley.)

Hook a chalk line in the center of the flow line at the bottom of the rolled roofing. Pull it taut to the center of the flow line up at the ridge. Pop the chalk line. This is the line for the bend in the metal valley.

When you felt in the main roof, continue the felt on to the far side of the rolled roofing up the valley. Cut the runs of felt just short of the flow line of the valley. Cut the felt so that the chalk line for the flow line of the valley isn't covered by the felt. Leave the runs of felt loose over the rolled roofing up the valley. You want to avoid extra nails in this critical valley area. If it's a breezy day, weight the loose ends of felt down with bundles. (If it's suddenly very windy, bring the felt across one run at a time and shingle up that run of felt until you're ready for the next run of felt.) In Figure 23, we have already run rolled roofing up the valley, stapled in, and cut off four runs of felt at the flow line of the valley.

When you felt in the wing roof, run the felt across the rolled roofing up the valley. Cut that felt at the flow line and leave those ends loose too. At this point, you can see the backup roofing function of the felt. Any rain hitting the main roof or the wing roof will roll down the felts to the rolled roofing in the valley and be carried on beyond the fascia board.

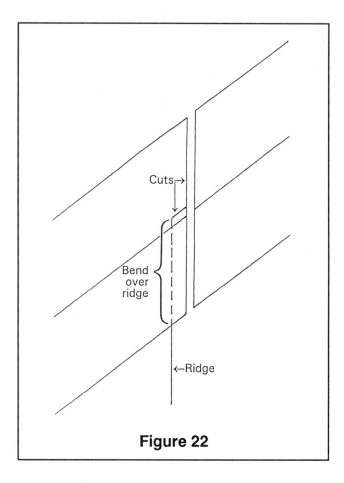

Figure 22

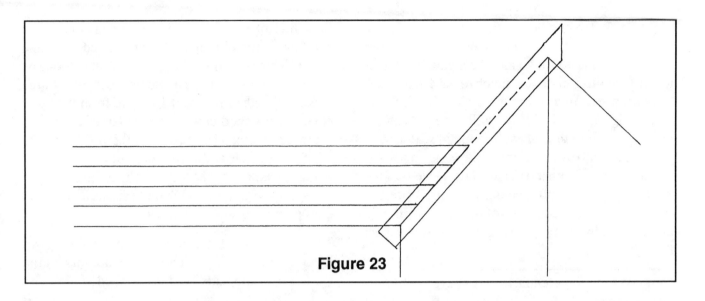

Figure 23

You lay a roof from the rake toward the valley. You want maximum tab strength at the rake, but you are going to cut the shingles over the metal valley at an angle (wherever the chalk line hits). Therefore the width and size of tabs at the valley is going to be different for each course of shingles. Set the verticals to give maximum tab length at the rakes. On the main roof, you will be working from the rake to the valley, and you will be doing the same thing on the wing.

The ridge of the wing roof may be like the wing in Figure 23 — it does not meet at the ridge of the main roof. The upper courses of shingles on the main roof will continue on over the wing without interruption. Therefore, you need to set the base line and offset line on the main roof so that the tabs have the maximum length possible at both rakes of the main roof.

If you run out of rolled roofing partway up the valley, lap the next roll 6 inches down over the roll you just finished and keep going up the valley. The lap will be slightly raised, but the metal valley and shingles will bridge over it and it won't be visible.

If you want to be absolutely sure everything is smooth, use whole rolls on the front valleys or the valleys that are most visible from the ground. Use shorter sections of rolled roofing with any overlapping joints on valleys that can't be seen as well or can't be seen at all.

A metal valley can be aluminum, enamel-coated aluminum, copper, or galvanized tin. Don't use galvanized metal. It's a pain to keep painting it. Copper is durable but extremely expensive. You again have the option of the .027 gauge or .019 gauge aluminum. Go with the heavier .027 gauge. The .027 aluminum will last beyond the life expectancy of the new roof and beyond the life of the overlay twenty years from now, too.

You should have the required sections of metal valley fabricated before you tear off. You don't want to have the roof open and have to run off looking for a sheet metal shop.

Aluminum is bent to the required shape in a metal brake, which is a machine that locks the metal in place and bends it along a straightedge.

I recommend (and urge) you to use 24-inch coil stock for the valleys. This comes in 50-foot coils. If you have access to an 8-foot brake, uncoil 8 feet and run a nail along the straight edge of a "square" to scribe (scratch) where you want to cut the aluminum. Use metal shears to cut the line you scribed. Measure 12 inches across the end of the aluminum and dimple it lightly at the 12-inch measurement with a punch or nail. Mark the aluminum this way at both ends. You are now ready to make the bend for the flow line in a metal brake.

You can cut a piece of cardboard to show the approximate size of the angle of the valley. Make sure you cut the cardboard so it fits when held in a position *vertical to the valley*. One comment about the angle that needs to go into the metal: it is better to have too little angle (too much toward being flat) bent in the metal than too much angle. With an angle that is slightly too small, you will have to push the metal down into the valley. The metal will tend to press down into the flow line of the valley once it is nailed in place. With too much angle, the legs of the metal will try to pull against the nails. With too much angle, the metal sides of the valley will bow up and away instead of pressing snugly down against the rolled roofing.

Nail: It takes some extra work to have the aluminum fabricated. I found that roofers who steered their customers away from metal valleys or refused to do a metal valley didn't own, or have access to, a metal brake.

You want each section of metal valley to overlap the one below it by 6 inches. If an 8-foot brake is used, and you have a valley that is 22 feet long, you have an ideal situation. You will have two overlaps of 6 inches each plus a 6-inch overlap of the metal over the ridge to the other side. You will need the metal section to overhang the fascia by 1 inch. To get the two lower corners to overhang 1 inch, the flow line will have to overhang 3 to 6 inches, depending on the slopes of the roof. Let's add another 6 inches to be on the safe side.

22' + 6" + 6" + 6" + 6" = 24', so 3 8-foot sections are exactly right.

However, suppose the valley is 19 feet and an 8-foot brake is the maximum length available. The nicest looking way to divide the metal is to have equal-sized sections of metal showing up a valley. In this example, we still need the 6-inch overlap over the fascia, two 6-inch overlapping joints, and the 6-inch overlap at the ridge:

19' + 6" + 6" + 6" + 6" = 21", so 3 7-foot sections are right.

The Building Code only calls for use of a 12-inch coil stock to form metal valleys. With a 12-inch stock, you will have a maximum 6-inch leg from the flow line of the valley. The Code also calls for bending an inch lip up and over both edges of the valley. This should be done with a brake, too. The bend for the flow line is made up the middle of the metal. Then the lip is bent 1 inch in from the outer edge and crimped down onto the leg of the valley using the brake. The reason for the lips along the edge is to catch and turn back windblown water being forced out of the valley. The effective leg of the valley with the 12-inch coil (minus an inch for the lip on each side) is 5 inches.

The Code also calls for the valley to be nailed down by driving the nails just outside the lips and letting the heads of the nails hold the bent edge of the lip down tight. You wouldn't want to drive a nail *through* these lips because water could seep down the nail shaft and ultimately cause problems.

Nail: I have seen roofers who did their 12-inch metal valleys right, and fastened the valleys by clamping the bend for the lips under the nailheads. They then laid their shingles and drove nails all through the metal valley. The company may have had the more experienced roofer on a crew do the valleys and special work and then had someone who only knew how to pound shingles do everything else. The result was a valley that leaked — a lot.

If you use a 24-inch coil stock, you are going to have a 12-inch leg of metal from the flow line of the valley. Beneath the metal, you are going to have a 20-inch leg of rolled roofing extending from the flow line. I don't recommend the protective lip at the edge of 20-inch metal valleys. If you have the free use of a brake and want to put a lip along each edge of the valley, that's fine. If you're paying a sheet metal shop, the protective lips aren't worth the extra price on the extra wide 24-inch metal.

You may notice that the flow line of the bare wood sheathing at the valley waves and shifts. This isn't a problem unless the flow line in the carpentry is extremely crooked. Set the lowest section of metal valley. Keep the bend for the flow line centered top and bottom on the chalk line you popped down the flow line of the rolled roofing. The metal goes over the loose ends of the runs of No. 15 felt coming across the main roof and wing roof. The metal valley will lock these felts down against the rolled roofing. Hold down one side of the metal valley, keeping the flow line firmly in position. Nail the top corner of the metal of the side you are holding in position. *Don't drive the nails all the way home* until you have all the metal valley sections in place. Raise the bottom of the metal valley slightly to sight under it, and set the bend for the flow line directly over the chalk line. Nail the same side at the bottom of the valley. Now go to the other side of the valley and, holding that side down, nail it from the bottom, working up to the top.

Now lap the second metal section down 6 inches over the first, and set the flow line bend of the second piece over the flow line bend at the top of the first piece. Nail the bottom corner of the second piece. Check the alignment of the bend by sighting underneath to the chalk line. (Or actually run the chalk line itself up from the bottom of the flow line of the first piece of metal valley.) Shift the top or the second piece of metal until this line is perfectly straight.

Repeat the process with the third section and let it overlap the ridge by 6 inches. Now get on the ground and sight up the valley. Is the flow line perfectly straight? It should be. If it's not, something has slipped. Since the nails aren't driven home, you can pull the metal loose easily. One thing: If you pull a nail loose and shift the metal, go ahead and move the metal clear and drive the nail back in the hole in the rolled roofing. The material will seal around the nail. After you have shifted the metal, use a new nail to secure it. Now finish nailing the metal sections in position. A nail every foot along the edges is sufficient. Valleys aren't subject to violent wind uplift the way rakes and caps are.

Measure vertically up the main roof from the edge of valley. Mark the courses $10^1/_2$", 5", 5", etc. If you tore off the main roof in stages and did the valley last, drive nails in the felt at the tops of the shingles where you stopped laying the roof. Hook a nail clip and pull a chalk line across the new course marks. Pop course lines over to the flow line of the valley.

Lay shingles completely across the flow line. As you continue each course, remember to nail down the ends of the shingles you left loose. Stop nailing the shingles down when you reach the edge of the metal. Just let the shingles lie across the valley.

Occasionally, you are going to lay a shingle that just barely reaches out onto the metal valley. When this happens, cut off the last tab of the last full shingle you laid. Start to cut at the center of the keyway. Don't try to make the cut exactly straight up the shingle; otherwise you will have to come back and trim the cut again. Angle the cut back just slightly toward the center of the shingle. Place another whole shingle in the place of the cut tab and nail the end of the whole shingle in place, keeping the nails out of the metal. This step eliminates small, weak slivers of shingles at the valley.

Lay the main roof on up until the valley is completely overlapped with shingles. At the bottom of the valley, use a tape to measure back into the main roof 3 inches. Hold the tape against the main roof on a line *perpendicular to the flow line*. Notch the bottom shingles at 3 inches and hook a chalk line in the notch. Sight an imaginary line that is 11 inches down (6" cap + 5" course) from the ridge line of the wing to the corresponding course you just laid on the main roof. Gradually trim a notch in the exposed portion of shingle beneath that imaginary line. Keep trimming the shingle until the end of the notch is 3 inches (perpendicular) from the flow line of the metal valley beneath. You get this measurement by running a tape through the notch until the end of the tape is at the flow line underneath the overlapping shingles. Hold the chalk line at the end of the 3-inch notch and pop the line.

Starting at the top, cut slowly and smoothly down this chalk line. Keep the hook blade perpendicular to the face of the metal underneath. Figure 24 shows how a finished cut will look on one side of a metal valley.

Trim the remnants to make cap pieces. Cut the cap pieces over a section of old roof. You don't want any slashes or scuffs in the new shingles. Store the cap pieces and throw the scrap away. Sweep down the new roof and valley if a lot of grit has accumulated.

Now finish roofing the wing and repeat the process of cutting the valley.

HORIZONTAL COURSES ON PARTIAL ROOFS

Tear off the other side of the main roof beyond the wing. If the main roof extends for some distance on the other side of the wing, tear off from the far rake back to within 4 feet of the valley on the far side of the wing. Felt in as you did before, overlapping the felt toward the old valley temporarily left in place.

Now you need to check and see if the fascia locations on the main roof are the same on both sides of the wing. Select a course that will extend uninterrupted from the main roof beyond the wing. There

are two ways to do this. You can use the first method if the main roof is square and has a straight bottom (or fascia) extending beyond both sides of the wing you need to cross. Measure from the fascia to the top edge of the course you will carry across the top of the wing. Now take the same measurement from the far fascia up the rake of the section of main roof on the other side of the wing. Make a mark for the top of the course at the far rake and drive a nail at the mark. Hook the chalk line and pop a line to the top of the course you will carry across. Go to the rake you started the main roof from and sight down the top of the course of shingles. Does the chalk line continue on a straight line with the completed course of shingles? If the chalk line is straight with the course, you are in good shape. Double-check the distance from the top of the course to the ridge and from the chalk line at the far rake up to the ridge. If the distance to the ridge is the same, the main roof

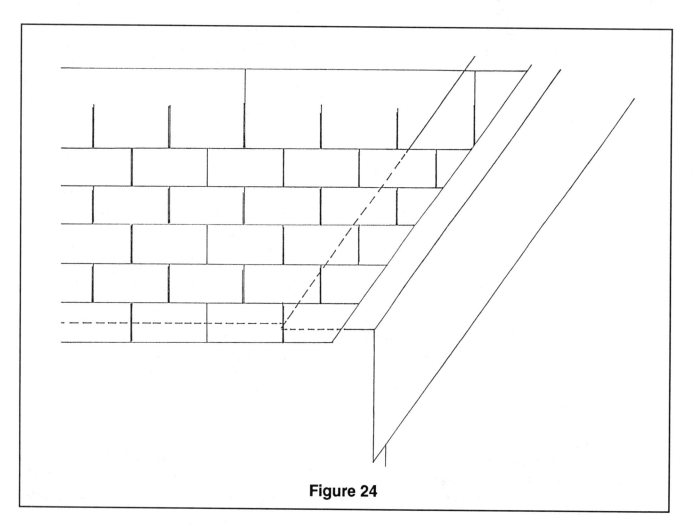

Figure 24

is square. You can mark the horizontal courses on up from the fascia board on the other side of the wing in the normal manner, and you know they will meet properly with the courses coming over the wing. Pop chalk lines for the horizontal courses. When you get above the top of the wing roof, extend a chalk line from the measurements or nails along the far rake back to the top of the shingles of the courses coming over the top of the wing roof.

If the main roof goes down to a point (no fascia) on the far side of the wing, you will extend the course line across the top of the wing. Drive a nail at the top of the course, keeping the nail back 10 or 15 feet from the last shingle you laid in the course. Hook the chalk line to the nail and line it up with the chalk line for the course to come across. If you have already laid the shingles to the wing, pull the chalk line tight over the wing, keeping it lined up with the top of the course of shingles you are carrying over. Pop the line. Go to the original rake and sight along the top of the course. Is the chalk line straight with the course? Good! You're ready to go.

If the chalk line moved up or dropped from the line the completed course was following, the main roof is not square. The fascia on the partial roof on the other side of the wing may be set differently than the fascia on the first portion of the main roof. In that case, set the chalk line straight with the top of the course of shingle you are going to carry over the wing. That's right, I finally said, "Eyeball it." Pop a new chalk line for the top of the course. Check it from the far rake and if it looks straight, measure from this final course line down the partial roof to set horizontal courses every 5 inches.

If the fascia on the partial roof is off just a few inches, you will have to make adjustments in the height of several courses of shingles. For an explanation of this see Chapter 17, where I show how to adjust courses as you lay up to a ridge.

If the partial main roof on the other side of the wing drops down to a point in the far valley of the wing, you have no fascia from which to measure horizontal courses. That's no problem. Again, in a case like that, you would measure 5-inch courses down the roof from the top of the course you carried across the wing.

CARRYING BASE AND OFFSET LINES BEYOND A WING

Now the only problem is the base and offset lines. Nail the courses on over the wing roof. The first course to come uninterrupted over the wing has to be nailed high. "Nailing high" means nailing the shingle not just below the self-sealing strip but up at the very top of the shingle itself. This will let you slide a lower course in below it. Nail this course and extend it on to a point between the wing and the far rake. Leave the end of the shingle loose. Bring the next courses up across the roof, stopping them at the same location. Leave all the ends loose. Nail above the ends of the top two courses of shingles. Measure the distance from the rake to the ends of these two courses. Take the same measurements at the fascia board and mark them. If the bottom of the partial roof is a point in a valley, pull the measurements out from the rake farther down the roof and mark them. Pop a chalk line at these two marks, and you have the new base and offset lines for the partial roof beyond the wing.

As discussed earlier, you don't want two consecutive courses meeting at the same vertical. So you want the shingles that will slide in below the course you nailed high to fall on the vertical line opposite the line for the course which is in place and nailed high. So you need to count odd/even down the roof. Remember to start counting *at the top* of the shingle of the course you nailed high. *Remember, the tops of two courses are under the shingle you nailed high.* Because the shingle you nailed high is 12 inches tall, the next course down will be 5 inches below and the second course down will be 10 inches below the top of the course you nailed high. If the end of that shingle is on the new offset line (closest to the far rake), the top of the next shingle down will be on the base line. Count the courses down: offset/base, offset/base, offset/base. When you get to where the first course is at the fascia, let's say the count ends on "O" for the offset line. Stop a minute!

Remember the starter course. Count one extra time for the starter course and end up on "B", the base line. The starter course begins at "B". The first course directly over the upside-down starter shingle starts on "O", the offset line.

With this system of counting alternating courses you end up either right or wrong. The chances are 50/50. I strongly recommend you lay a shingle (but don't nail it) for the starter course, put the first course shingle on top of it and continue placing one shingle on each vertical for each course up the roof. When you reach the course coming across the top of the wing, the shingle for that course should butt right into the end of the course. If the end of the loose shingle you lay to check the count falls on the base line, and the shingle it is supposed to butt into is on the offset line, you know there is a mistake. Check everything again. If you have the shingles loose-laid with the starter course in place too, trust the actual shingles and reverse the shingles all the way up. If they match properly after reversing them, nail that starter course in place and start the final shingling of the partial roof. The horizontal and vertical alignment will meet perfectly, and it will look as if the lines continued uninterrupted across the main roof despite the major obstruction of the wing roof. See Figure 25 for the sketch of counting down the roof.

FITTING SHINGLES AT TRIANGULAR BOTTOM OF PARTIAL ROOF

Many brands of shingle have cuts at the top of the shingle exactly over the middle of the tab (6 inches from the center of each keyway). The cuts are there to help you with situations like this. Find the notch in the top of the first shingle and bump the end of the second shingle against the side of the notch. This should give you a perfect 6-inch offset.

The second way to do this, if the shingles aren't notched in the top, is to measure over 6 inches from the side of one key to the corresponding side of the key of the next shingle up the roof. Keep piecing and nailing like this until you get the courses up to the base and offset lines you popped earlier. Then just

lay the courses on up and tie them in. Be sure to nail down the course you nailed high and the ends you left loose.

CAP OVER THE TOP OF A METAL VALLEY

Read Chapter 17 about the treatment of a ridge and caps. Many roofers will carry the cuts for the sides of the metal valley all the way up and over the ridge to meet the side cuts for the valleys on the opposite side of the wing. They mastic the cuts they made in the metal and just leave it at that. It's an approved method, but I like to carry my caps plus one course from the wing on across the top of the valley and weave the ends of the shingles in under the tabs of the shingles on the main roof.

Figure 26 shows how to cut the top of the aluminum valley. Trim the corner off the overlap over the ridge of the wing so that the cut is 6 inches over and parallel to the ridge of the wing. Then cut back in a direction perpendicular to that cut 6 inches to the flow line bend. This will let the metal drape down into the valley on the opposite side. Nail down the top of the metal from the lesser seen valley first, then overlap it with the metal from the most visible valley. Nail the overlaps over the ridge. You will probably have to nail the ends of the metal out across the flow line of the rolled roofing to make the metal stay down. Nail the metal legs up the main roof.

Caulk or mastic the exposed cuts you made in the aluminum valley where it goes over the ridge. You have double-overlapped the shingles going over the ridge of the wing. Carry the top course of shingles on both sides across the valley and tie the ends in neatly under the tabs of the course closest to that same elevation on the main roof. If a course overlapping the valley is going to fall just short of the flow line, cut a tab off the last full shingle and add a new whole shingle where you removed the last tab. The full shingle will reach across the valley and eliminate a sliver of a tab. Start laying the caps at the far end of the wing from the valley. Carry the caps across the valley to the main roof. Shape and mold these cap pieces and nail them as needed. Water will run off the top of these shingles and on down the valley from there, so these nails won't cause any problem. Raise the tab of the corresponding shingle on the main roof and nail the covered end of the last cap piece up under that tab.

Trim the top of the metal valleys to suit your taste. I usually went one course down from the cap piece and trimmed across the bottom of that shingle.

Carrying the shingles across the way I have shown you gives the top of the valley added protection. Mastic under the lower edge of the shingle course across the top of the metal valley to keep the wind from trying to raise these shingles. Then mastic the bottom edge of the tab down to the top of the shingle you cut off. If the corners of any cap pieces are tending to stick up where you tied them into the main roof, mastic under those corners and weight them down with a bundle for a couple of days until the mastic sets. The last step is to run a bead of mastic under the cut edges of the shingles to bond the shingles to the aluminum valley.

Metal valleys are the hardest to do, but they are the most durable and give the nicest finish to your home.

DOUBLE-WEAVE VALLEYS

If you decide against metal valleys, the double-weave valley is the next best choice. The rolled roofing and No. 15 felt are done the same way on a double-weave valley as for a metal valley. *On all valleys, you keep the nails as far away from the flow line as possible.* You double-weave a valley by laying alternate courses of shingle from each roof completely across the valley and across each other. The sequence is as follows:

1. Lay the starter course from the main roof across the valley and on up the wing roof approximately 1½ feet. Press the shingle snugly into the flow line and nail the far end of the shingle to the wing. Keep nails away from the immediate area of the flow line of the valley.

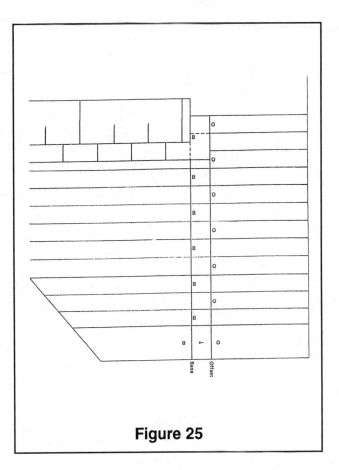

Figure 25

2. Lay the starter course from the wing roof across the starter course from the main roof and across the flow line of the valley. Hold the shingle snugly in the flow line and nail the far end of the shingle into the main roof.

3. Lay the first course from the main roof across the starter course from the wing roof and across the flow line of the valley. Hold the shingle snugly in the flow line and nail the far end of the shingle to the wing roof.

4. Lay the first course from the wing roof across the first course from the main roof and across the flow line of the valley. Hold the shingle snugly in the flow line and nail the far end of the shingle to the main roof.

5. Lay the second course from the main roof across the first course from the wing roof and across the flow line of the valley. Hold the shingle snugly in the flow line and nail the far end of the shingle to the wing roof.

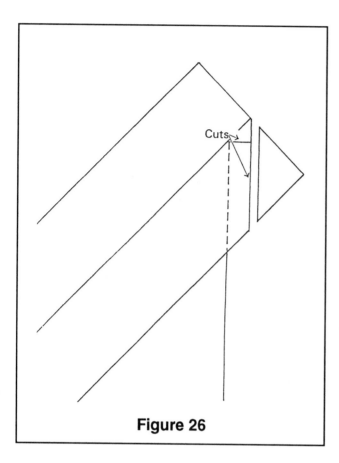

Cuts

Figure 26

6. Lay the second course from the wing roof across the valley and across the second course from the main roof and hold and nail.

7. Lay the third course from the main roof across the valley and the second course from the wing roof and hold and nail.

8. Lay the third course from the wing roof, etc. You get the picture.

Once the double-weave is done, don't step down in the valley. You might crack or loosen some of the shingles.

A double-weave is a good fix if you are overlaying the roof and have a metal valley that has been leaking. It is also good if the old metal valley is galvanized and rusting. The double-weave will take the old valley completely out of service.

"CALIFORNIA CUT" OR SINGLE CUT VALLEYS

Again, the underlayment under this valley is the same as for a metal valley or a double-weave.

Lay the shingles on the main roof across the valley approximately 1½ feet beyond the flow line of the valley. Push the shingles down snug in the valley and nail the ends of the shingles. Now lay the wing roof overlapping those shingles through the flow line of the valley. Pop a chalk on the wing roof shingles. The line is 2 inches *from* and *parallel to* the flow line of the valley and located on the wing roof itself. Cut the shingles from the wing roof along this chalk line. Mastic down the cut edge of the wing roof shingles and the California Cut valley is done.

This leaves only the shingles on the main roof in the flow line of the valley. The main roof shingles do all the work and get all the wear at the flow line. This is the basic weakness of the California Cut.

Nail: The California Cut is an acceptable installation. It's the quickest, easiest, cheapest, and flimsiest of the three choices. I think it's bad enough to rate a "Nail."

VALLEYS ON AN OVERLAY

To reuse an existing metal valley, "nest" the courses to the old shingles and let the courses run across the valley. Keep the nails in the new shingles away from the edge of the metal valley underneath the old roof. Notch ¼ to ⅜ inch beyond the cut edge of the old shingles and pop a chalk line. Cut the new shingles and mastic them down to the old shingles. Covering the old shingles by ¼ inch takes them completely out of service.

A double-weave and California Cut are done the same way they would be on a tear-off.

Here is the bitumen for the rear valley. I had to cut the tops of both of them so they would lie flat. Note: Felt is rolled to the flow line (bottom of the "V") of the front valley.

Rain caught us the evening before. Here are the short valleys weighed down with bundles — no leaks.

The modified bitumen is in place, and the felt has been rolled to the flow line. Doug helped me install copper valleys over the felt. One side benefit of metal valleys is that the metal helps lock the felt in place in the event of a sudden storm with high winds. Note: There is a 6-inch overlap where the two sections of metal valley meet.

The roof of the front wing is basically done, and I finish-cut the shingles at the copper valley. I am laying shingles on up the main roof. Note where I rough-cut the lower shingles on the main roof at the end of the previous day.

I have just finished laying the shingles that lap completely across the valley from the main roof.

Doug is carrying a course on the upper roof over the valleys tying in the lower roof. He is nailing the course high (top of the shingle, staying above the keys).

I counted, recounted, and then laid the shingles in rough just to be sure my count down the roof was right.

Start of double weaving—first the course from the lower roof goes across the valley to the upper roof. Then the starter and first course combination from the upper roof go across the valley. Then the next course from the lower roof crosses to the upper roof.

Upper roof to lower roof, then lower roof to upper roof.

Upper roof to lower roof; see, it's straightforward.

Bringing both front and rear valleys up together.

Closer ...

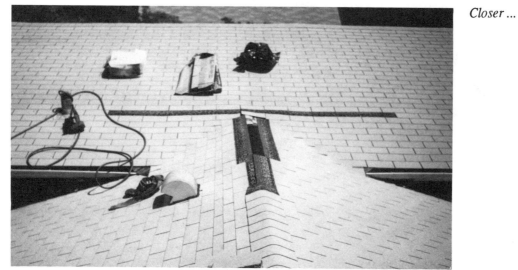

The double weaving is done.

17.
Caps and Ridge Vents

The ridge of a roof is another trouble-prone area. In a severe high wind (such as a hurricane), the wind buffeting the side of a house takes the line of least resistance and races up and over the ridge of the roof. The result is that the shingles along the ridge are hit with a greater force and experience more severe pressure than the shingles anywhere else on the roof. The treatment of the ridge and laying of the cap pieces require special attention.

On a complete tear-off or brand new home, everything should be double-overlapped on the ridge, starting with the roofing felt. Lap 1 foot of the felt from the rear roof over the ridge and staple it to the front sheathing. Lap 1 foot of the felt from the front roof and staple it to the felt underlayment and sheathing on the rear roof.

You will overlap the tops of the courses of shingles the same way you do the felt. Let me point something out.

If your front roof and rear roof have the exact same height, start the starter course on the front roof at the *base* line and start the starter course on the rear roof at the *offset* line. If you use the same measurement for the base line on the front and rear roofs, and you start both starter courses at the base line, and come up the same number of courses to the ridge, then the butted joints for the shingles you double-overlap will meet exactly. If this happens, the butted joints are open every 3 feet along the ridge and the opening goes straight down to the roofing felt below.

It's true that you are going to cap over the ridge and the caps will cover the matching joints, so it will be watertight. It won't be a problem until the cap pieces get old and crack. Then it can be a problem.

CAPS

We have talked in earlier chapters about trimming the overhangs over fascia boards, trimming into valleys, trimming into chimneys, and so on. Save all the whole tabs (full 12-inch widths) that you trim off and work them up as cap pieces. Cap pieces are made by laying the cut piece with the full tab on the old shingles with the shingle face up and the tab up the roof away from you. Hook the blade at the point where the straight side of the key ends and the curve for the top arc of the key begins. Cut the shingle from the key back at a slight angle toward the center of the tab. Do the same kind of cut from the other key or the half-key. You should end up with a whole 12-inch tab with straight sides and the top cut slightly back toward the center of the tab. None of the curved arc at the top of the key should remain. This is called "hip cutting" the cap pieces.

Nail: Very few contractors hip cut their cap pieces. A straight cut from the center of the key to the top of the shingle is all that is required to make caps. The problem with a straight cut is that the curved top of the key always hangs out from under the finished edge of the caps and the straight cut is never really straight. The hip cut only leaves the machine-cut edge of the key showing, so the line of caps is perfectly straight.

The reason a lot of contractors don't hip cut their cap pieces is that for every key you have to make *two* hip cuts instead of *one* straight cut. In other words, the hip cut requires twice as much cutting.

As you cut the caps, stack them finished side up and directly on top of each other. Remember the self-sealing asphalt strips can bond the cap pieces together unless the caps are stacked so that the protective tape on the back of the shingles still works. Make sure that the pile is uniform and neat. Drape the pile of caps across the ridge of the roof. It will make it easier to lay them if the heat of the day has already basically molded them in the shape you want them to take. If you aren't going to use the caps that same day, still cut and stack the caps every so often as you go along, and lock the stack of cap pieces down with an unbroken (still sealed in the paper wrapping) bundle of shingles. You don't want a sudden high wind flinging those things at you. They act like giant flying razor blades in a high wind.

Here is an example of when to cut caps. When you finish cutting one side of a metal valley, stack all the trimmed pieces that are big enough to cut for caps. Go through the small pile and cut the caps, throwing away the remaining small trimmings and trash. Stack and store the cap pieces, and the job is cleaned up for you to cut the next side of the metal valley. This makes for less mess to slip on and less mess to cause a mistake.

If you are roofing in the winter, keep the caps warm. If you let the caps get cold, the asphalt gets brittle. If you bend cold caps over the ridge, you can crack them. A cracked cap is nearly worthless. In cold weather, keep the caps inside in a warm spot, preferably near the furnace. Don't bring them outside until you are actually ready to use them. Then leave them in a stack so they retain their heat until you are ready to use them.

Double-overlapping the shingles in cold weather can also crack the shingles. If you are doing this

work in the winter, try to save the double-overlapping and capping until the warmest part of the day.

Nail: Many contractors do not overlap their shingles or their roofing felt. They depend completely on the caps to protect the ridge of the roof. The only positive reason that I have heard for not double overlapping the shingles is that the caps lie flatter if there aren't any shingles under them. This is a lame excuse for building a weak roof. We did many re-roofing jobs, and many roofs we overlaid were originally done with no double overlap of the shingles beneath the caps. Some people waited too long to reroof and had suffered real water damage to the interior of their homes before reroofing.

Now you may need to make a decision about the upper courses. The tabs on the shingles are 12 inches wide, so the cap pieces you've cut are 12 inches wide. That means when you nail them over the ridge of the roof, you will have basically a 6-inch leg of the cap coming down each side. In a perfect world, you lay the 5-inch courses right on up the roof and the top of the last course would overlap the ridge by an inch. That would leave the top of the keys 6 inches down from the ridge, and you could lay the caps in perfect position right to the top of the keys. Unfortunately, the world is an imperfect place.

For instance, let's say you have 14 inches between the bottom of the second course down the roof and the peak of the ridge. You could fill this space with another 5-inch course. Because the shingles are 12 inches high, that final course of shingles will overlap the ridge by 3 inches. (14" space - 5" for next course = 9 " to ridge from bottom of shingle. 12" shingle - 9" distance to ridge = 3" overlap.) This is a good overlap and should be easy to nail solidly in place. However, when you nail the caps on, the 6-inch leg going down the roof will only leave 3 inches of the final course showing. (9" distance to ridge - 6" leg on cap = 3")

The other option would be to reduce the final course to a 4-inch course. That also works. (14" distance to ridge - 4" final course = 10" to ridge. 12" shingle - 10" to ridge = 2" overlap) This may be a little more difficult to nail if the ridge isn't perfectly straight or there is a large gap in the sheathing at the ridge. However, when you nail the caps on, they leave 4 inches of the final course showing. (10" to ridge - 6" leg of cap = 4")

In this case, whether you want a 5-inch course, a 3-inch course, then a 6-inch cap; or a 5-inch course, a 4-inch course, then a 6-inch cap is up to you. However, the second combination is less noticeable.

Now that you've decided how you want to do the courses, the next problem you face is that you've run out of horizontal course lines. Let's say you decided to go with the 5"-4"-6" cap combination. Measure up the roof 4 inches from the bottom of the last course of shingles you laid. Drive a nail at 4 inches into the last tab along the edge of the roof and hook the chalk line on it. (I know. After you pop the line, you are going to have to replace that tab. When you get to the rest of the roof leave that last shingle off so the nail hole won't create extra work. I apologize; it was easier to show you why to leave that shingle off than to go into a detailed explanation.) Play out the chalk line across the roof and measure up 4 inches from the bottom of the last course on the far end of the roof. Pull the line tight beneath the tape at the 4-inch mark and pop the line. Go back and remove the shingle or tab I made you ruin by driving the nail. Replace that shingle or tab.

Lay the shingles, keeping the bottom edge of the shingles right on the new course line across the tabs. Don't lay the last shingle on the front roof. You need to pop a line for the caps.

I always overlapped the shingles from the rear or least visible roof first. Then I overlapped the shingles from the front. The front shingles "bridged" over the nailed "tops" of the overlapped shingles from the rear. When I nailed on my cap pieces, the caps covered the nailed "tops" of the front shingles where they overlapped to the back roof. This sequence gave the front, or more visible roof, a nicer fit and finish.

Putting the last few courses in and double overlapping takes a lot more time than you would think just

reading about it. If darkness or exhaustion is overtaking you, the ridge of the roof is substantially sealed even without the cap pieces.

Center a cap piece over the ridge of the roof where you left that last front shingle off. (The cap piece goes up and over the ridge of the roof.) Once the cap piece is centered over the ridge, drive a nail even with the bottom of the cap piece. Hook a chalk line to the nail and gently reel it out to the far rake. Center a second cap piece over the ridge at the far rake and pull a chalk line tight along the bottom of the cap piece.

Remember, some brands have a small slit at the top of the shingle right in the middle of the tab, which is now the cap piece. Use it as a guide to center the two cap pieces. Pop the line. You want the line you will use to lay the caps on the front roof. The line in the front will be perfectly straight. If there is any sagging or variation due to an uneven ridge line, let it be to the rear where it won't be seen.

The prevailing wind in our area is from the west. Most of our storms seem to blow from the west. On a house with a ridge line that ran east and west, I always started capping on the east end. This left the upraised or exposed end of the cap piece facing the direction *away* from the prevailing winds. The next most likely direction of severe winds for us is from the south, so on a ridge with a north/south orientation, I would start capping on the north end.

Center the cap piece over the ridge of the roof with the tab going up and over the ridge. Line up what would have been the bottom of the tab with the cuts on the rakes, and then line up what was formerly the side of the key with the chalk line for the caps. Nail the front leg of the cap just toward the exposed surface from the self-sealing strip. This action tends to roll up the self-sealing strip, so the exposed edge of the next cap will seal down more completely to it. Mold the cap down securely and nail the rear leg down. Now nail the two corners at what was the top of the shingle. You have the first cap piece in place.

Now center the second cap piece over the ridge, leaving the 5-inch tab of the first piece exposed. Nail it just as you did the first piece.

The Building Code calls for each cap piece to have four fasteners, whether you use nails or staples. I put four nails in each cap piece. Then when I overlapped the next piece, the two nails closest to the exposed tab portion of the second piece also went through the first cap piece. This means each cap piece was held in place with six nails.

Continue laying caps across the roof until you get to the far end. Now if you keep going, you are going to have the upper end of the shingle exposed at the end, instead of the tab portion with the finished grit on it. Keep going the way you have and let the top of the last cap stick out over the cut rakes by 2 inches. Nail the cap piece in place. Cut off the overhanging portion even with the rake cuts. Now cut off the top 4 inches of a cap piece and turn this piece to run the other way, keeping the bottom of the tab portion even with the rake cuts. Nail this piece down with two nails, again placing them just toward the tab from the self-sealing strip.

Now we need a "locking cap." Cut the exposed part of the tab off a cap piece. Start a cut where the top of the key would have been and cut straight across the tab to where the top of the opposite key would have been. You should end up with the exposed grit-covered piece measuring 5" x 12". Center this rectangular piece over the overlapping tops of the last two cap pieces you laid. It will cover them nicely and give the roof a finished look. Nail down each corner of this rectangular piece. (Use aluminum nails.)

Cut another rectangular piece and go to the far end of the ridge. Center it over the exposed edge of the second cap piece and nail the rectangle down at all four corners. You should use aluminum nails on these, but a galvanized nail is all right if you put a dab of mastic or caulk over each nailhead. If you have been able to get a silicone caulk with a color similar to the shingles, that's even better. The caps are now securely locked at each end of the ridge.

If you do the caps this way, and your house goes through a hurricane, you will find that a tremendous amount of other damage will have occurred before the storm starts to bother the caps on the roof. It will take a history-making storm to damage caps done the way we did these.

Nail: I have never seen a roofer in our area put more than two fasteners in each cap piece. Also, most of them just continue their cap pieces on to the end rather than turning the last one and installing "locking caps."

RIDGE VENTS

In recent years, a large number of homeowners have installed ridge vents when they reroofed their homes. To install a ridge vent, you basically cut a 1- to 2-inch slot through the shingles and sheathing along the ridge of the roof. You cover the slot with a ridge vent unit, which is, in effect, a roof over a roof. Convection currents through the attic carry the hot air out through the slot you have cut in the ridge of the roof and through the tiny louvered vents on the underside of the top section of the ridge vent unit.

Ridge vents carry a lot of heat out of the attic, and they are reasonably durable. However, I'm not convinced that they will hold up well in severe storms.

You install a ridge vent by continuing up the roof just as if you were going to double overlap the shingles. However, now you will want to measure how far the legs of the particular ridge vent will go down the roof from the ridge. Adjust the last couple of courses into the ridge accordingly, but remember, at the ends of the roof you are going back to the double overlap and caps.

Let's say you want to use ridge vents on a section of roof where the ridge is 30 feet long. Ridge vents come in 10-foot sections, and they can be cut with a hacksaw. You decide you want to go with the installation of 20 feet of ridge vent.

There are plastic, vinyl, and aluminum ridge vents on the market. (You know how I feel about plastic on the roof.) Go with the aluminum. You will need to buy two 10-foot sections of ridge vent, one connector for tieing the two 10-foot sections together, and two rubber ends to seal up the ends of the ridge vents.

Lay the shingles up and over from the back to the front, but don't nail down the overlap. The slot in the sheathing is supposed to be cut down right at the end of the ridge vent so the vent unit can sit right down beside the caps. On a 30-foot roof with a 20-foot ridge, the computations are:

30' length - 20' of ridge vent = 10' length of ridge that will need cap pieces

You generally are going to want the ridge vent centered on a roof, so you divide the remaining capped length in half:

10'/2 = 5'

Measure 5 feet from the end of the roof: this is where you will start the ridge vent.

As I said, I'm not sure I trust ridge vents. A straight cut, even with plenty of caulk and mastic, is most likely going to cause a problem eventually. We should leave another foot of double overlapped shingle under the end of the ridge vent just in case. So mark 6 feet from each end of the ridge and cut the 18 feet (30'- 6' - 6' = 18') of shingles down 1 inch from the ridge line on both the front and back roofs. (A chalk line will help keep this cut straight.) Overlap and nail the 6 feet of shingles located at each end of the ridge cap placement.

Now lay the courses up the front roof. Again, cut the 18 feet where the ridge cap will be and then overlap the 6 feet of shingles at each end onto the rear and nail them down to form the double overlap on the 6-foot end sections.

Many roofs have a gap in the planks or plywood at the ridge of the roof. If the gap already there is, say, ³/₄ inch, don't bother to widen it. If the plywood is jammed right up tight along the ridge, you will need a circular saw to cut a couple of strips out of both sides. Set the saw so that the blade will cut down through the plywood sheathing without going through the 2 x 4 rafters beneath. *Wear protective goggles.* Chances are, you are going to hit the edge of a shingle or maybe even a hidden nail. Stuff will

fly back in your face like a small explosion. Don't be a hero and think it can't happen to you.

Keep a careful check on the depth the blade is cutting. Cutting across all the roof rafters and making them collapse is only funny if a professional stuntman does it in a movie. You should have set the depth of the saw so that the plywood slots you cut are still hanging by a thread. Knock the cut slots out with a hammer. Sweep and remove the trimmings, sawdust, and grit that have accumulated.

Now run a heavy bead of mastic on the shingles below the slot you just cut. Make sure you keep this heavy bead well within the area that the base or legs of the ridge vent will rest on. Use a generous amount of mastic, but not so much that it will squeeze out down the roof below the ridge vent. Now go back to the measurement — 5 feet from the end of the ridge — and run a bead straight up and over the double overlapped shingles, tying it in with the parallel bead you ran on the back of the roof. Insert the rubber end in the ridge vent and set the first section of ridge vent down in the mastic.

Now butt the second 10-foot section against the leading end of this first section. Mark where the far end stops. It will be 5 feet from the far end of the roof ridge. Finish laying a bead of mastic along both the front and back of the slot and then lay a bead up and over beneath where the rubber end will be located. Set the second section of ridge vent in mastic with its end firmly butted to the first section.

Come back to where you butted the ends of the two pieces of ridge vent together and run a bead of mastic up and over the top of the crack between them. Lay the connector piece up and over them, but don't nail it yet.

Sight down the ridge vents and get them lined up straight down the ridge line. Now nail the corner nearest you, but don't drive any of the nails down tight. (A bag of twisted shank aluminum nails should have come with each section of ridge vent.) A nail set is handy when installing most designs of ridge vent. The set will help you keep from hitting the raised flange on the outside of the unit and bending it down. Stand up and sight along the ridge again to make sure everything stayed straight. Nail the far end of that same 10-foot section where the two sections join. Go back and sight down it again to make sure it's still straight. Now nail the far end of the second 10-foot section. Sight from there and see if both units are still straight. Go back to the center and, keeping the two sections lined up, nail down the end of the second 10-foot section. Nail the center nails down snug and nail the one end of the joint piece down. Drive the nails snug at both ends. Repeat the same process on the other side of the roof. When you have the entire length of both sides snugged down tight, drive a nail in every pre-drilled nail hole in the ridge vent units. Go back to the ends and seal along the base of the rubber end pieces with mastic. (I usually covered the whole end piece with caulk.)

Nail: Many roofers install ridge vents without using mastic at all. The unit's legs going down the roof are not very long. If rain blows in under the units, it is going to drop into the home through the open slot in the roofing and sheathing. Also, I have seen cases where other roofers only drove nails in every other or every third pre-drilled nail hole and used smooth shaft galvanized roofing nails to install the unit. The first blast of high wind will take those ridge vents.

Nail: Some roofers have been known to install "lookalike" ridge vents. They never cut the slot in the shingles and sheathing. The installation looks like a ridge vent, but it doesn't ventilate anything.

I don't like ridge vents. However, this method of installing them should keep them in place and keep you high and dry. If your hands get covered with mastic, kerosene or diesel fuel will take it off.

THE ROOF IS NOT SQUARE

I have run into cases where the roof was not square. It's a good idea to stop popping horizontal lines 3 or 4 feet down from the ridge and measure along each rake from the last line you popped to the center of the ridge. Let's say you are down eight courses from the ridge line, and the distance at the left rake is 45

inches. This will put you just right coming up to the ridge. Let's be a little extreme and say the distance on the right rake is 49 inches. The right end of the courses will come into the ridge 4 inches lower than the left end.

Double-check the measurements for the 5-inch courses at the left and right rakes. Make sure the chalk lines are popped correctly. Check across the final chalk line you popped. Does it fall down from the ridge continuously across the roof? We decide the carpentry on the roof is definitely out of square.

We have eight courses to make up a 4-inch shortfall on the right rake. It will weaken the roof and possibly expose some of the black grit portion of the upper part of our shingles if we try to increase the height of the courses on the right. The only thing to do is decrease our measurement for the courses going up the left rake. 4"/8 courses = 1/2" per course. We have to decrease our left rake measurements for the last eight courses to 4 1/2 inches, while we leave the measurement for the courses at our right rake at 5 inches. Someone looking up from the ground probably won't detect the 1/2-inch variation in these last eight courses, and the top course of shingles will be even with the ridge line and the cap pieces.

CAPS ON AN OVERLAY

Overlay the old roof until the new courses on both legs of the roof are up within a few courses of the old caps. Tear off the old caps. If the courses beneath the cap are cut at the ridge or are crooked coming to the ridge, you should consider tearing off those couple of courses, too. Lay up any new courses you need at the ridge and cap the ridge as described earlier in this chapter.

Nail: On an overlay, one of the first steps many roofers take is to remove the cap pieces. It is neater and easier to remove the caps first. However, if the shingle courses at the ridge aren't double-over-lapped, the roof is open and waiting for rain. Too many roofers don't have a roll of felt or tarps on an overlay job. They don't have a way to seal that ridge if a sudden storm hits.

It's a better idea to tear off the old caps when the new shingles are up to the caps on both sides of the roof. Then you know the caps are going to be replaced immediately.

Hip cut cap pieces.

Old caps using straight cuts. These courses and caps are much straighter than a lot I have seen. This roof was done before I bought the house.

The finished roof!

The author's son, ten-year-old Daniel T. Chiles, proves that with the right tools and instruction, anyone can roof a house!

18.
Cleanup

Cleaning up the job is more than just a matter of aesthetics. Gently sweep down the roof as you finish each section. You won't be able to see all of the flaws in the shingles or the "fish mouths" while you are standing on the roof. Have someone on the ground sight up the slope of the roof and direct you to the raised shingles. You can help the person on the ground by holding your hammer vertically down on the roof and standing to one side. Lift the raised tab and remove any trash or loose grit and drive down the nail if it isn't seated flush with the surface of the shingle. Remove all scrap pieces so they don't stick down to or get stepped down into your new shingles. Don't wait until you have done the whole roof to do the detail cleanup. If you wait, you may discover that the tabs are already sealed down, and some scraps have stuck to and stained your new shingles.

GUTTERS AND DOWNSPOUTS

Clean the gutters and downspouts well. Clean the gutters using a narrow horsehair brush and a piece of step flashing. Flush the gutters out with a garden hose. The hose gets all the dirt out of the gutter and guarantees that the downspouts are clear. Scoop up any grit, nails, and scraps at the bottom of the downspout.

If the downspouts are blocked with leaves or dirt, there is one very effective way to clean them out. Stuff the nozzle of the hose down the top of the downspout and jam rags in around the hose to seal the top of the downspout tight. Then have someone on the ground turn the hose on full force. You will get wet and spattered with dirt, but the water pressure will blow the crud out of the bottom of the downspout. If water pressure won't clear the blockage, hold the hose and rags with one hand and beat on the side of the downspout with the heel of your other hand. That, combined with the water pressure, should shake things loose. If that fails, try a plumbing snake. If that fails, get ready to take the downspout apart.

Nail: Some roofers will clean your gutters down far enough that you can't see any trash sticking up from the ground. The first heavy rain shifts that trash to the outlet into the downspout and your gutters flood. Some blocked gutters back the water up over the fascia and into the house.

Nail: I have heard roofers say they cleaned up any shingle scraps or nails that they caused to roll into the gutter. They left anything else where it was. If the gutter was loaded with dirt and rotten leaves, it wasn't their problem. They had cleaned up their own mess and left the gutters the way they found them. If I told them what I thought, they said, "I never said I'd clean out no gutters."

Nail: Several years ago, a guy I was training did an amazingly fast job of cleaning some heavily loaded gutters. I checked his work. The gutters looked good. When I washed out the first gutter, water spilled out everywhere. He had jammed as much of the shingle scrap and nails as he could down the downspouts rather than taking the trouble to lift the trash from the gutter into the bucket. I had to take a few of the downspouts apart, but I finally got everything cleaned out.

Tighten the spikes and ferrules on your gutter. Use the flat side of your hammer to drive the spikes down snug. It will save you dinging the gutter. If your roof was a tear-off and there are scrap and nails behind your gutter, partially loosen the entire gutter by tapping out on the inside top of the gutter with the side of your hammer. Shake the gutter vigorously so the trash will fall out from behind it. You may have to run a screwdriver up in behind the gutter to loosen the roofing nails that have wedged into the wood of your fascia. Now tighten the spikes.

GROUND CLEANUP

Clean the ground thoroughly. If you had to throw

tear-off on the ground, you will never clean up all of the loose grit. The grit will eventually work into the ground. However, you don't want to step on an old piece of self-sealing strip on a sweltering summer day and then walk across the light carpet in your living room.

Pick up every nail you can find. A yard rake will turn up nails you missed the first time. There is also a tool called a *nail rake*, which you might want to rent. It is a heavy-duty 30-inch bar magnet suspended on wheels. Mine worked great at picking up galvanized nails, but of course it won't help you with aluminum or copper nails.

Nail: A poor roofer will often ignore a loose nail on the ground or try to stomp a nail down into the grass or dirt. The roofing nails from your old roof are galvanized. They aren't going to rust away. The blade speed on a rotary power mower is approximately 3,000 rpm. When your mower blade sucks up that nail and fires it, the top of your foot or shin may suffer the most painful "nail" of all.

19.
Super-Strength Roofing

You can give your roof super strength in resisting wind damage. If you live in a hurricane-prone area, on top of a mountain, or in some other location where you regularly experience severe winds, you might want to consider the handful of ways you can give your new roof super strength.

SHINGLES

Stay away from the standard twenty-year shingle. Use a heavier duty, self-sealing, twenty-five-year, three-tab shingle. Most manufacturers' thirty-year and thirty-five-year shingles are dimensional shingles with the rough surface that simulates a wood shake roof. I don't have scientific data to prove it, but it's logical that the rough surface gives extremely high winds a place to begin tearing off the roof. I suspect that the wind resistance caused by the rough surface more than offsets the strength of the extra thickness of the thirty- and thirty-five-year shingles.

MASTIC FOR STRENGTH

Mastic the lower edge of your plywood or the bottom plank of the sheathing (just above the gutter) and roll the felt out over top of the mastic. If you do this, you should consider buying a 5-gallon bucket of mastic and a small trowel.

Another method is to mastic the bottom edge of your starter course to the felt just above the gutter. Leave regular gaps in the mastic between the felt

and the back of the starter shingles. Remember, the felt acts as a backup roof, and someday moisture may have to drain off over the felt but underneath your starter course. Regularly spaced gaps in the mastic will let the moisture run down to the gutter.

Some roofers tie the bottom tabs down another way. They cut all of the tabs off the shingles in their starter course and lay the starter course right side up. The top of the shingles in a starter course with the 5-inch tabs cut off follows a course line $5^{1}/_{2}$ inches above the face of the fascia board. The cut edge of the shingle and the self-sealing strip then hang out over the fascia $1^{1}/_{2}$ inches. The first course is then laid using a course line $10^{1}/_{2}$ inches above the face of the fascia. The first course seals down to the self-sealing strip on the cut shingles of the starter course. I think it is much easier and much neater to just do it the way I showed you originally and then follow the next step (below).

Mastic the tabs of your first course of shingles down to the starter course. Just put a dab of mastic under each tab. You don't want the mastic squishing out around the keys. Using mastic on the felt, starter course, and tabs of your first course cements each component at the bottom edge of the roof to the sheathing.

FELT

You can use No. 30 felt for the "field" in place of the No. 15 felt. Or you can double your felt as I showed you in the low slope configuration. If the roof has a pitch of 4-12 or more, I don't recommend that you go to the 4-inch courses if you are using a standard three-tab shingle. When you go to the 4-inch course, you raise the self-sealing strip 1 inch higher beneath the bottom edge of the next course of shingles. Then instead of having 1/2 inch to 3/4 inch of tab lapping down below the self-sealing strip, you would have a 1 1/2-inch to 1 3/4-inch overlap not sealed down. These longer unsealed bottoms of the tabs give the wind a place to start lifting and this defeats what you are trying to do.

However, some dimensional shingles have the self-sealing strip under the bottom edge of the tabs. The bottom edge of these tabs is going to seal down to whatever is underneath no matter what measurement you use for your courses. If you have decided to use this type of shingle despite its greater wind resistance, the bottom edge of the tabs will seal down the same as it would if you used 5-inch courses. (Remember to increase your shingle order by 20%).

NAILING

Six-nail the shingles instead of four-nailing them. In other words, you will nail at each end of the shingle then drive nails on both sides of the two center keys. This increases the resistance to wind uplift by 50%. Incidentally, when you nail the next course, the nails coming through from the shingles above will give you twelve nails holding each shingle instead of the normal eight nails.

You can also use 1 1/2-inch roofing nails, double-nail the rakes, and four-nail the cap pieces and use locking caps.

FLASHING

Instead of the 5 x 7 step flashing, use the larger 9 x 12 step flashing, which has a 3-inch vertical leg and

a 6-inch horizontal leg on a 12-inch long piece of step. If 9 x 12 step is not available, use some other similar large dimensioned step.

ACCESSORIES

For a stronger, more wind-resistant roof, don't add pot vents, ridge vents, power ventilators, skylights, or other penetrations through your roof. Severe hurricane winds can tear these items away.

Drip edge and roof edge can add strenth if they are properly installed. If you don't use roof edge, reinforce the rake edge with twenty-five-year three-tab shingles running up the rakes as shown in this book.

Use rolled roofing or modified bitumen around brick or masonry chimneys and as underlayment for valleys as I have shown.

Stay with heavier gauge metal for skirt flashing, valleys, and any other items you fabricate.

Install metal or double weave valleys. Thicken the mastic bead under the trimmed ends of the shingles to the metal valley. Stay away from the California Cut.

Bed all vent flashings, and, if you must have them, pot vents and power ventilators, in mastic. Mastic the shingles down to the base plates and then caulk the finished edges of the shingles with silicone.

A standard roof that is properly done will withstand a real beating from the wind. This chapter contains several new steps that will make your roof super strong. As you can imagine, these additional steps will make the work much more tedious and time consuming.

Remember that nothing manmade lasts forever. I can only assure you of one thing. If you follow these extra strength recommendations, yours will be the last shingle roof in your area to blow off when a "category five" hurricane hits.

20.
Warranties

Most reputable contractors include a standard one-year to five-year warranty covering their workmanship. The statement of this warranty should be included in the contract. Any problems on a self-sealing shingle roof are generally going to show up the first year. All the tabs will seal down tight long before a year has passed. A year takes the roof through a complete cycle of seasons. If you are talking to an outstanding contractor who offers a one-year warranty on his workmanship, please realize that it is a reasonable warranty.

Nail: It doesn't matter what the warranty says if previous customers tell you they can't get this contractor to come back and take care of a problem.

Nail: It also doesn't matter what the warranty says if you discover that this contractor is in the habit of changing his phone number every month, and you are never able to reach him again.

I received "call backs" occasionally. Once the leak was in the customer's new aluminum siding, which had been installed after I completed his roof. I had to refer him to his siding company. Another time a two-year build-up of pine needles and pine cones was forcing water out of an aluminum valley I had reused on an overlay. I cleaned the valleys and gutters. A third customer's front roof had a tab that delaminated. I replaced the customer's tab, and there have been no more problems on this six-year-old roof. When I got the calls, I dropped what I was doing and went immediately to correct the problem. All three calls came in after the customer's warranty had already expired. A reputable roofer will fight to keep his reputation.

When you give the roofer the final payment for the roof, he should give you the original of the manufacturer's warranty. This form should give your name and address and the address of the home that was reroofed, if the project wasn't on your residence. The form should show the name and address of the roofing contractor. It should give the style of shingle used and its color number. There usually isn't a space for the blend number, but I always included it with the color number. The contractor should send a copy of the form to the manufacturer so the manufacturer can keep the information on file.

The manufacturer's warranty is going to make exceptions for hurricanes, hail, falling trees, lightning, vandalism, etc. These exceptions are reasonable.

Keep the contract, contractor's warranty form, and manufacturer's warranty in a secure file.

Nail: Many, many contractors in our area don't give the homeowner the warranty form. If they do give the homeowner the warranty form, they don't send a copy to the manufacturer. Even the better contractors don't follow up on this important item. Don't make the final payment (remember the "pay on completion — poc" part of the contract) until you receive the warranty form already filled out and signed. Companies cut the corner on filling out the warranty as a way to trim their overhead.

You should go ahead and send a copy of the form to the manufacturer yourself. Your roofer should have already done this for you, but let's not ass/u/me anything here at the end of the game.

If you have done the roof yourself, you will have to send a copy of the warranty form to the manufacturer. It is a good idea to insist on receipt of a blank warranty form from the supplier when you purchase the shingles. So many roofers are lax about these forms that suppliers aren't careful about keeping the forms available.

Do the manufacturers' warranties amount to anything? In 1984, '85, and '86 there was a rash of premature roof failures in our area. All the homes (and their roofs) were built in late 1973 and 1974. I couldn't understand it at first, but then I remembered the trouble I had with asphalt in 1973 and 1974 when I was a highway engineer. The oil companies made all kinds of alterations to their product when the 1973 Arab oil embargo cut availability and made prices skyrocket. Some of the stuff we were sent to use on the highways was a mess. By the mid-80's it was obvious that, back in 1973 and '74, the asphalt used to make fiberglass/asphalt roofing shingles was also changed by some of the oil companies or roofing manufacturers or both. To make a long story short, the subdivision developers who had protected their customers by sending in the roofing warranties made it possible for the owners of the affected homes to receive a refund of part of the cost of the replacement roof.

These shingles were laid straight up, and you can see where whole bundles failed.

21.
Other Roofing Materials

There are many types of roofing materials available in today's market. I have dealt with the fiberglass/asphalt shingle because it is durable, virtually maintenance free, and reasonably economical. If you reroof your own home, you can expect to save approximately half the cost of having it done by contract. You may decide to do what I did and use the money you saved on labor to pay for top quality thirty-year dimensional shingles and copper valleys and trim. Going to significantly higher quality levels with your materials definitely enhances the appearance and future marketability of your home. However, as with anything else, there are limits at which you reach the point of diminishing return.

SHAKE OR WOOD ROOFS

Let's say you are crazy about the appearance of a cedar shake (wood) roof. I'm with you: a cedar shake roof really looks good. Shakes look so good that, in our area, the contractors building large two-story homes routinely charge an additional $10,000 for a cedar shake roof. Shake roofs can last thirty or forty years if you buy premium (read that as "extra-expensive"), heavy-duty materials. However, the average shake roof will last about twenty years, and you will be doing some expensive patching toward the end of that time.

Shakes need to be laid on an open-gap plank roof. They need to be backed with rosin paper, which will let the underside of the shakes breathe and dry easily. If the shake is laid on plywood with roofing felt, the underside of the shakes tends to stay moist while the top side dries completely in the blazing sun. This uneven drying means it won't be too many years until the sides of a lot of the shakes have curled upward from the middle.

A shake roof should be torn off before it is replaced with a new shake roof.

For the do-it-yourselfer, a shake roof is unbelievably tedious. Instead of dealing with a bundle of 3' x 1' shingles, you are dealing with variable widths of wood that have to be laid so that each piece overlaps the piece in the course below it by a certain horizontal distance. That means fishing through the bundle of wood until you find the right piece. When you finally do find the right piece, you've got at least, *at least*, a one-in-ten chance that it will split when you nail it.

A shake roof should be fairly steep. The surface of a 5/12 or 6/12 shake roof is a whole lot slicker than fiberglass asphalt shingles at the same pitch. Plus, every now and then, one of the shakes you nailed simply splits and lets go when you step on it.

If you put a shake roof on your home, your insurance company will give you the same rating they would a pile of kindling wood. You can buy shakes that are treated with fire retardant, and they do have a higher fire rating — initially. Unfortunately, every rain and every snow washes away a little of the chemical. In several years, there won't be any significant difference in the treated or untreated shake. Fiberglass asphalt shingles carry a Class A fire rating. They will burn, but they don't burst into flame at the first spark. Go to shakes, and your homeowner's insurance premiums will transfix you with their breathtaking climb.

When you sell your home twelve years from now, with its twelve-year-old shake roof, a knowledgeable prospective buyer is going to count the shake roof as a definite liability.

Check out dimensional shingles available from your supplier or contractor. Keep up your search until you find a fiberglass asphalt shingle that gives you the look you like.

SLATE AND CLAY TILE

Many Victorian homes in the 1800s and early 1900s were built with slate roofs. Slate is an excellent material, and these roofs can last for eighty years and more with minimal or no maintenance. When you do need to maintain a slate roof, you will find that there are very few roofers around who deal with slate. If you get on a slate roof yourself, I predict that you will crack four pieces for every piece you patch. The roofer you finally find to help you will charge prices not very far behind those charged by a brain surgeon.

If you are thinking about tearing off your existing fiberglass asphalt shingle or tin roof and putting a slate roof on your home, it might be all right. It's OK, provided your home was originally designed and built to hold up the tremendous weight of a slate roof. This means that your home must be strong enough from the foundation, through the basement walls, through the exterior walls, through the interior bearing walls, and through the truss system and sheathing of the roof to hold up the large number of tons of slate. Do you know if any of the interior bearing walls has been altered or removed since the house was built? Has the truss system been altered by skylights, or have the trusses been changed to open up space for a room in the attic? If you think your home may have had a slate roof at one time and you want to restore it to slate, let an experienced building engineer check your house before you start laying the slate over your head.

Clay tile or orange Spanish tile roofs present the same weight problem that slate does.

STANDING SEAM METAL ROOFS

Tin roofs, or, more correctly, standing seam metal roofs, are fading rapidly in our area. We still see standing seam roofs done in copper over bay windows in storefronts or in front of upscale homes. It is extremely rare to see an entire roofing system done in standing seam.

Tin

A standing seam tin roof will last a long time, given proper maintenance. The problem is the maintenance; it needs to be painted every three to four years. You can let it go longer, but exposed tin corrodes. The cost of hiring painters every few years would have paid for the fiberglass asphalt shingles.

Copper

A standing seam copper roof will last for years. You buy copper by the pound now. It is rapidly approaching the price per pound of a medium cut of steak. Copper also has the nasty trait of staining any trim that it drips on. On my roof, I used copper on my valleys and trim, but that was enough for me. My *aluminum* gutters will catch any stains from the copper valleys.

Stainless Steel

Let me wander with you again. In 1976, I was the field engineer overseeing the restoration of a covered wooden bridge, which had been almost totally destroyed by a Halloween arsonist. Wooden bridges were always protected by slab-sided walls on both sides and a roof completely over the top. This 1894 bridge was 210 feet long and nearly 15 feet wide. Its original roof was cedar shake. The "powers that be" decided to put back a standing seam stainless steel roof. It would last virtually forever. The cost of the stainless steel roof itself (in 1976) was $28,000. The opening ceremony was wonderful. What had been lost was restored.

The next week some fool shot the roof full of holes with a thirty aught six. We patched the bullet holes with stainless steel nuts, bolts and washers, and rubber donut washers inside and out. It's still that way today as far as I know. So much for lasting forever.

CHOICES

We all have choices to make. Your home has a roof on it now, and when you have finished all your hard work or have paid your contractor, it will have a roof on it again. It might make you mad when it happens, but no matter what you put on top, or how beautiful your work is, some people aren't going to notice anything different. Because I know this to be so, I have tried to steer you toward what I consider to be the most attractive and most serviceable roof for your money. If you really want people's attention, take the $10,000 you saved by *not* having a contractor put on a shake roof and use it as the down payment on a new red Corvette. That, everyone will notice and talk about.

22.
Routine Maintenance

A fiberglass asphalt roof does the best when it is disturbed the least. The shingles do "cure out" after they have been laid a few months. The grit is less likely to scuff off after the roof has aged a little. However, every time you walk on it, a little bit of grit lets go and a little bit of cracking might occur at the edges of the tabs. It's not disastrous and it's not usually noticeable. But you don't want to get in the habit of taking a daily walk on your new shingle roof.

When you do get up there, don't step or stand on your caps. You have them double-overlapped and securely fastened, but why ask for trouble. (I got one reroofing job because the youngster in that house watched parades while standing on the peak of the lower roof. Several buckets were strategically placed in their family room when I got there.)

Don't step in your double weave valleys and risk cracking a shingle. Don't step in your metal valleys; your print might stay in the metal as a reminder.

CLEANING GUTTERS

I have already mentioned the roof leak caused by pine needles blocking a valley. Leaks can also occur because of clogged gutters and/or downspouts. If the roof is not too steep, clean the gutters from the roof. If you hose the gutters out, try to keep the shingles above the gutters as dry as possible. If you get careless on wet shingles, they will throw you in a heartbeat. If you're standing above the gutter and you slip, there is no room to recover.

BLOCKED DOWNSPOUTS

You can unclog a downspout by pushing a garden hose down through the outlet from the gutter into the top of the downspout. Pack around the hose with rags and hold everything in place. Have someone on the ground turn the hose on full blast. The pressure should clear the downspout. If it doesn't clear, beat on the side of the downspout with the heel of your hand. That, combined with the water pressure, should break the clog loose. If that fails, try the same procedure from the bottom of the downspout. Sometimes the upward pressure works.

If water pressure fails, you will have to disassemble the downspout to clear it. If the sections are held together with rivets, you can drill out the rivets, clear the blockage, and re-rivet the downspout together.

GUTTER SCREENS, DOWNSPOUT STRAINERS

Gutter screens cover the tops of your gutters. They run down the gutters lengthwise and fit beneath the lip on the front of the gutters and beneath the overhanging bottom edge of shingles. Screens do some good and may keep your gutters open an extra year. You will pay for it in aggravation and nicked hands when you do finally have to clean the gutters. I've seen enough of gutter screens to last me forever. But I shouldn't complain. I got some roofing jobs because the customers had small trees growing up through their gutter screens.

TAB REPLACEMENT

I had one tab delaminate and go bad in ninety roofs. Sometimes a tab is damaged by a limb. To replace a tab, lift up the tabs above the damaged tab. You can usually raise a tab with your fingers. However, if a tab is sealed down tight, slide a bricklayer's mortar trowel (or similar flat blade) under the tab to break it loose. Pull the nails out of the tabs above the damaged tab. (This will let you slide the replacement tab in place.) Once the tabs above the damaged tab are raised, pull the nails from the damaged tab. Use a hook blade to cut the damaged tab loose. Start the cut as near the top of the damaged shingle as you can, and cut down to the center of the keys of the damaged tab. You will probably end up pulling the loose pieces free with your fingers. Slide the new tab in place, nail it, and renail the old shingle you left in place above the damaged tab. You should mastic down all the disturbed tabs.

NURSING AN AGED ROOF

Shingles crack and tear and can get nail pops toward the end of their useful life. Don't try to replace tabs in a very old roof. Drive the nails back down flush and mastic over these nails. Also, mastic the cracks in the shingles. Check the caps and mastic any cracks that are developing directly over the ridge. If leaking is occurring along a wall or at the corner of a chimney, careful application of mastic may carry the roof a little longer. Cracks in a cement chimney cap can also be sealed with mastic. If you have a collection of these kinds of problems, you know you have waited too long to replace your roof.

COATINGS

There are some coatings available for shingle roofs. These might buy you some time, but you can probably guess how I feel about them. The coatings generally are going to make it harder to tear your roof off when the inevitable time comes. You are going to pay for the coating, and you are going to pay a premium again if you hire someone to tear off and reroof your home.

WHAT TO EXPECT

If you keep trash from accumulating on your roof and keep branches cut back so they don't rub the roof, your fiberglass/asphalt shingle roof should be virtually maintenance free. Keep the gutters and downspouts clear and the spikes driven in snug.

OVERLAYS

Remember, if your house only has one roof on it, you can overlay it. An overlay works best when the old shingles are still lying flat and haven't started to crunch underfoot like cornflakes and crumble into tiny pieces. It is better to overlay two years too early than to wait one year too long and have to do a complete tear-off.

CHECKLIST FOR A NEW ROOF

You need a new roof when:

1. The shingles are brittle and crumbling underfoot.
2. The shingles are cracking, and you see extensive nail pops.
3. You find pieces of shingle in your yard.
4. You can see extensive areas that look swollen at the corners, or the shingles have obvious "bear claw" curling.
5. You have a leak or leaks showing inside the house, and your roof is getting toward the end of its life. Often, your roof will leak for awhile before it shows up inside. Check your attic. Don't be surprised if you find rotted sheathing.
6. Sections of shingles have lost most of their protective grit.
7. You notice recurring leaking around potential trouble spots, such as valleys, chimneys, or vent flashing.

There may be some other troubles peculiar to your specific roof that make you think you need to replace it. If you have that feeling, follow your instincts. You are now knowledgeable about roofing, and you are probably right to think you need a new roof.

23.
Summary of Sequence of Work

The following are step-by-step condensed lists of the sequence of work for (1) a tear-off, and (2) an overlay.

SEQUENCE OF A TEAR-OFF

1. Order materials and have all materials on the job before tearing off the first shingle.

2. Check to make sure you received the correct materials, including the type and quality of shingles and the color.

3. Check blend numbers on the ends of the bundles to make sure they are all the same. If you have a few odd-numbered bundles, move them to the least visible roof now.

4. Cut away any overhanging limbs.

5. Paint any walls or trim above a section of roof.

6. Repair or replace cement cap on brick/stone chimney.

7. Wire brush and spray (high heat paint) metal chimneys.

8. Rough clean the gutters.

9. Tear off the roof of a free-standing garage or shed and reroof it first to gain the experience.

10. Start on the section of roof on the house that is farthest from your truck or trash pile.

11. Start tearing off on the other side of the ridge of the roof you want to remove. Work up the other side to tear off the two courses, felt and caps.

12. Standing on the ridge, get your "shingle eater" or spade under the top of the felt and shingles on the section of roof you are working on. Work your way back and forth across the roof, raising the shingles and working them free. Stay up on the clean sheathing to have the safest footing. Keep the trash cleaned up and sweep the sheathing and old sections of roof to keep the grit off.

13. "Roll" up the old roof as you tear off along the gutter.

14. Remove old roofing nails or drive them down flush. Remove any old scrap pieces of roofing.

15. Replace or add sheathing nails (8d nails) and drive down any that are pulled partway out.

16. Replace rotted sheathing and repair or replace sagging or rotted rafters.

17. Rough clean the gutters again.

18. Take a break and drink lots of water!

19. (Optional) Install aluminum drip edge along the fascia or lower edge of the roof. Mark and notch the short leg of the drip edge and bend the notches up to make room for the spikes and ferrules of the gutter.

20. (Optional) Roll out and adhere "ice shield" to the sheathing along the lower edge of the roof. The ice shield should cover the upper leg of the aluminum drip edge.

21. Roll the first run of No. 15 felt out along the lower edge of the roof. The lower edge of the felt should extend over the lower edge of the sheathing. If you used drip edge, the felt should extend over the edge of the drip edge over the gutter.

22. If you use ice shield, cover it with the first run of felt so you don't have to work on the slick surface.

23. Start the felt with a "pivot point" of staples. Keep the staples set horizontally for maximum strength. Bump the roll on the far rake to get a mark for a straight cut. Turn the roll and come back across the roof, making sure you maintain a 2-inch vertical overlap. When the roll ends, the new roll should start with a 6-inch horizontal overlap of the old roll.

24. Shape all-lead flashing to fit down over the plumbing vents. In the event of a sudden storm, you can mastic a "donut" around the pipe and seat the flashing down into place.

25. Slice and patch any large wrinkles in the felt.

26. Overlap the ridge of the roof with the felt, making sure the felt extends down over the tops of the old shingles on the roof on the other side.

27. If rain is threatening, move the piles of shingles from the ridge down to the old section of roof on the side opposite the one you are working on. Now overlap the felts over the ridge.

28. Pop verticals — base line and offset. Get them toward the center of the roof. Stay away from piles of stocked shingles so you can pop the lines all the way to the ridge.

29. Pop horizontal lines.

30. Stock the roof by spacing the bundles of shingles on the roof. Drop one end to break the bundles. Naturally, you want the finished side up and the tabs pointing down the roof.

31. (Optional) If you are using an air gun, tie your hose off to a plumbing vent, chimney or a pile of bundles so you aren't pulling against the weight of the hose all day. Stock a coil of nails above each bundle.

32. Begin the "starter course" at the baseline. The tabs point up the roof and the top of the shingle hangs out over the fascia board or gutter. Carry the starter course completely across the roof.

33. Start the "first course" at the offset line and lay the shingles directly over top of the "starter course."

34. Lay your second course starting at the base line again.

35. Cut off some of the extra weight of shingle hanging over the rake. Trim the rake once this section of roof is completed.

36. Have a spotter on the ground direct you to any "fish mouths." Snug the nails flush with the shingle and remove any grit or scrap that is keeping the shingles from lying flat.

37. Get a good night's sleep.

SEQUENCE OF AN OVERLAY

1. See steps 1 through 8 on the above list.

2. Don't disturb the old caps until you have shingled both sides of the roof up close to the caps. Don't disturb the old plumbing vents until you are ready to replace them.

3. Notch the rakes top and bottom and pop a line. Cut off the old shingles overhanging the rakes.

4. Pop a line above the fascia and cut off the old shingles overhanging the fascia (gutter).

5. Measure 11 inches above the fascia and cut off the old shingles overhanging the fascia (gutter).

6. Measure $10\frac{1}{2}$ inches above the fascia and pop a line across the old shingles. This is the line for the top of the starter/first course.

7. Pop verticals — the base line and offset.

8. Lay the starter course shingles from the base line. The tabs point up the roof and the top of the shingle lies out over the fascia or above the gutter. Lay the starter course all the way across the roof.

9. Lay the first course directly over the starter course, beginning at the offset line.

10. Lay the second course beginning back at the base line.

11. Cut excess weight off the shingles at the rake. Trim the rake when the section of roof is finished.

12. Have a spotter on the ground direct you to any "fish mouths." Snug the nails flush with the shingle and remove any grit or scrap that is keeping the shingles from lying flat.

When the roof is done, be sure to send the warranty information to the manufacturer.

If you tore off a cedar shake roof, notify your insurance company so they can lower your premiums.

24.
Scaffolding

I have put off talking about scaffolding as long as possible. I do not like the thought of an inexperienced person on a roof steeper than 6/12. I include this chapter only because some of you may be forced to do your own roof, no matter how steep it is. If you have to do it, you should know how to do it as safely as possible.

We did one roofing job on two large townhouses that shared the same roof. The front of the roof was two stories up, had an 18/12 pitch, and rose 18 feet from the gutter to the caps. There were six protruding dormers with individual peaked roofs and metal valleys. In addition, the old roof was asbestos shingles so we had to wear very restrictive face masks while tearing the old roof off. One of my men quit and the remaining man had only done a couple of roofs before we tackled this monster. Frankly, I didn't want the project and bid it high in an effort to avoid it. I didn't know until we finished the job that the other roofing companies would not bid on it at all.

I always used adjustable roofing jacks that nail right into the shingle roof and can take up to a 2 x 10 scaffold board. You can adjust the support leg on the jack so that the board stays level on various pitches. On this monster roof, the adjustment was on its last notch.

You start the steep roof working from a ladder. It is extremely tedious, but lay the first three courses across the roof. Place the scaffold jacks no more than 8 feet apart (count eight keys). I routinely used a 16-foot scaffold board supported by three jacks. If the roof is especially steep or dangerous, put the jacks closer together and go to a 2 x 10 plank. On the monster roof, I had the jacks 5 to 6 feet apart and used unfinished (untrimmed) 2 x 10 planks.

Each scaffold jack is centered over the keys of the last course of shingles you laid. When you are starting a roof, you set the jacks on top of th ethird course. There are three nail slots in the upper part of the jack, and the bottom nail should be driven approximately 1/2 inch above the top of the key. Drive the remaining two nails at varying angles to keep the jack from jumping loose. Use 8d or 10d nails.

Nail: Some roofers hold their jacks in place with a couple of roofing nails. The results are sometimes disastrous. The nails should be galvanized, because they stay in place when the scaffold is removed.

You will notice that since the jack is over the key, the tab of the next course of shingles up the roof will cover the nails.

Lay the scaffold boards in position on the jacks. The arm that the board sits on has a vertical end piece to hold the board in place. That vertical end piece has a nail hole in it. Nail through the hole and into the scaffold board. You don't want your scaffold board jumping loose either.

When you overlap scaffold boards, make sure there is a jack supporting the overlap. Don't overlap a scaffold board onto another unsupported scaffold board. Charlie Chaplin made millions laugh doing that type of thing, but nobody will be filming you.

Drive a nail through both boards of the overlap. You don't want a board to slip.

Nail: Some roofers who do use jacks and scaffold planks instead of the 2 x 4 toe boards (chicken ladders) just lay the scaffold planks up there without nailing through the jacks into the planks.

Use good quality planks that are in good condition. Don't use painted planks. The paint gives the planks a slick surface and covers up flaws. You don't want to suddenly discover that the only thing between you and eternity is one weak knot hole.

The scaffold will support you and enough bundles to lay the roof up as far as you can reach from the scaffold. Stock the bundles on the scaffold plank and walk on the bundles too. The bundles are a tripping hazard. The overlaps of the scaffold planks are a tripping hazard. In addition, if you are using an air gun, the hose can get caught and unexpectedly jerk you. Move slowly and easily along the scaffold. (If you start to feel like an expert, get back on the ground fast!)

You will scaffold up the roof in stages. Work from your first scaffold to build the second stage of scaffold. Repeat the process on up the roof. When the roof is complete, you bring your scaffold down in stages, too. Remove the scaffold planks. Lift the tab covering a jack. Use your hammer to tap (sometimes pound) the base of the jack up the roof until it slides free of the three nails. Drive the three nails down flush and mastic over the heads. Press the covering tab down flat over the three nails.

TIP: If the jacks are going to stay in place and are going to get wet, they can leave a rust stain. Raise each jack up slightly and slide paper from a bundle wrapper between the bottom of the jack and your new shingles. The paper can protect your new shingles from rust for weeks if necessary.

I scaffolded everything right and still had one jack pull partially loose on me on a very steep roof. I sidestepped until another jack was supporting me and then I hustled down the ladder. It turned out that the sheathing underneath the jack was weak. The only reason the jack didn't pull completely loose was that I had varied the angles of the three nails holding the jack. I was only one story up, but it was a sickening feeling when that jack started to let go.

Nail: The shortcuts roofers take by not scaffolding or by just using "toe boards," warrant a Nail. If a roofer is doing stupid stuff and risking his own safety, you can guess what else he is doing as he lays your roof.

If you have hired an uninsured roofer, and he gets hurt, he has no choice but to sue you and your homeowner's insurance. If an insured roofer gets hurt because of something dumb, he nails the entire industry because all insured roofing businesses end up paying higher insurance premiums. The cost of the increased premiums is passed on to the customer.

25.
In Closing

If you are contracting the work for your new roof, you will *not* find a contractor who will do all these things exactly the way I have described them. Each roofing company develops its own characteristic methods. My goal is to prepare you to recognize and appreciate top quality workmanship when you check your contractor's references. I want to make sure you are ready to negotiate in what can be a ruthless game. The next steps are yours.

If you are overlaying or tearing off the roof yourself, you know exactly how to do a top quality job. You need to plan your strategy and your sequence of work. Don't be overly concerned about the weather. Just schedule your work around the weather forecast, and keep a "what if" factor in your plans. You know how to handle sudden storms. When you reroof your home using this book, you will have a top quality roof with an outstanding appearance.

(You will also save a considerable sum of money.)

When you finish the roof, you will probably hear the kind of comment that was always music to my ears. More than once, prospective customers told me, "I want you to do my roof for me. I don't know why, but your roofs always look so much better than any others I see being done."

I have done everything I can to show you what to do and also to explain why you should do it that way. Now, it's time to start talking to suppliers and their previous customers.

Whether you are contracting or reroofing your home yourself, I hope it is a pleasant experience for you and you achieve outstanding results. If my direction and insight helped you, it was my pleasure.

Now, go to it!

We converted our two-car garage to a master suite. The triangular roof sections create problems, because water doesn't run straight down the side triangles. The front triangle had to be done first so the shingles on the side triangles would have the "top" overlap at the ridges.

Nearing completion. To do this roof, I marked the arrows for the verticals using a large carpenter's square held against the drip edge. Everything done here was an application of basic roofing techniques. And you can do it, too!

Glossary

BASE LINE — This is one of the two vertical chalk lines you pop up the center of your roof. You can shift the layout of your shingles on your sheathing by the number of inches you subtract from an even multiple of 3 feet.

BITUMEN — This is a type of pliable but solid bituminous-based material. See MODIFIED BITUMEN.

BREAKING THE BUNDLE — This is the act of dropping one end of the bundle of roofing shingles down while holding the other end. The impact on the old roof or sheathing causes the shingles to suddenly shift loose from each other in the bundle, and it makes it easier to pull the shingles free one at a time.

CALIFORNIA CUT — This is the least desirable of the three valley types. The shingles from one roof are laid completely through the valley and cut 1 foot beyond the flow line of the valley. The shingles from the second roof are laid across the valley, a chalk line is popped 1 to 2 inches short of the flow line, and the shingles on the second roof are cut on the chalk line. The shingles from the first roof carry all of the water down the flow line.

CAP — The shingles laid lengthwise atop the ridge of the roof.

CAP PIECES — These are the individually cut tabs of shingles that are nailed over the ridge of the roof.

CUTTING THE RAKES — The process of marking and cutting the new shingles that will provide the protective overhang over the rake of the roof.

DEAD LOAD — The weight that your rafters and truss system always have on them. For example, the weight of the sheathing, felt, and shingles are permanent loads and are therefore the "dead load."

DIMENSIONAL SHINGLES — These shingles are not just flat on the roof. They have random "three dimensional" tabs that rise from the roof surface to simulate the finish on a shake roof.

DOMED VENT — This term is generally reserved for the larger covered attic vents that don't have a fan.

DOUBLE WEAVE — Valleys can be done by bringing the sections of roof on both sides up together. The course from one roof is laid across the valley. Then the course from the second roof is laid across the valley. This repeated back and forth overlapping has a weaving effect and gives you a double thickness of shingles through the valley — a "double weave" valley.

DRIP EDGE — Technically this is the bottom edge of the roof. More commonly it refers to the optional aluminum trim piece that goes over the sheathing and beneath the felt.

DUTCH ROOF — A slang term for a HIP ROOF.

FASCIA — The fascia is the piece of wooden trim covering the cut ends of the rafters at the bottom edge of the roof.

FELT — An asphalt-impregnated material that is laid on top of the sheathing before shingles are laid.

FLASHING — Metal trim around plumbing vents, chimneys, skylights, walls, etc. See STEP FLASHING and VENT FLASHING.

GABLE — On the majority of roofs, the end walls continue straight up to a peak beneath the ridge of a roof. The triangular area of the wall between the front and rear fascia and the ridge is the gable.

GAMBREL ROOF — The gambrel roof has a steep pitch on the lower segment of the roof, and then the section from the lower section up to the ridge has a lesser pitch. For example, the pitch of the lower section of the front roof may be an extremely steep 18/12 and then change to a 4/12 for the lower portion of the roof on up to the ridge. The gables on a gambrel roof continue to rise vertically from the walls on the end of the building.

GOOSENECK — A type of vent or cowling that comes up out of the roof then bends back down toward the roof so the opening faces downward and is protected from precipitation.

HIP ROOF — The front and rear roof rise to the ridge. There are no gables at the ends of these roofs. Additional triangular roof sections rise from the tops of the end walls to meet the ridge.

HIP-CUTTING — Shingles are cut to form cap pieces using a cut that starts at the side of the key and comes straight up the solid portion of the shingle with the cut angling slightly back toward the center of the tab. Hip cutting keeps loose ends from showing when the caps are laid.

ICE DAMMING — This problem generally occurs when a wet snow has done a little melting during the day and then frozen solid at night. The moisture flows down the roof and saturates the snow above the gutters. This slushy snow then freezes solid in and above the gutter. The next day the heat escaping from the attic and the sun cause the snow on the roof to melt and run down until water starts standing on the roof behind the ice dam. Water will then come in under your shingles, and it will usually leak into your walls or interior.

"ICE SHIELD" — This brand name is becoming generic for materials that adhere along the lower edge of the roof sheathing to prevent water penetration due to ice damming.

KEYS — The notches made in a three-tab shingle.

KEYWAY — The straight alignment of the notches in shingles.

LEAN-TO — A lower roof which rises to tie into a wall, in in effect "leaning" against the wall.

LIVE LOAD — The active or variable load on the rafters or truss assembly. Rain water, snow and ice, and the force from the wind are all examples of a live load on a roof.

MANSARD ROOF — The front and rear roofs have a steeply pitched lower segment and less steeply pitched upper segment. The triangular roofs rising above the end walls also have a more steeply pitched lower segment and a less steeply pitched upper segment.

MASTIC — Asphalt-based roofing cement.

METAL BRAKE — A metal bending machine that clamps one section of metal and bends the remainder over a straightedge.

MODIFIED BITUMEN — A type of rolled roofing composed of bituminous materials. Recommended for use as an underlayment beneath metal or double weave valleys.

NAIL GUN — A pneumatic (air) nailer. It holds coils of roofing nails which vary up to 1½ inches in length.

NESTING — When overlaying an existing roof, the tops of the new courses of shingles slide up against the bottom edges of the tabs of the next old course of shingles up the roof. The new shingles nest against the old ones.

OFFSET LINE — The second vertical line which controls the space and keyways of the courses of shingles which alternate from the base line courses.

PAPER — A generalized term for roofing felt. A derivation of "tar paper."

PEAK — Another word for the ridge, or top of the roof.

PITCH — The rise of the roof from horizontal.

PLUMBING VENT — A pipe that rises through the roof and ventilates a plumbing system to the outdoors.

POT VENT — A small attic vent with rounded or square cover. The round ones look like upside down pots on the roof.

POWER VENTILATOR — A domed venting unit which includes an electric fan.

RAKE — The edge of the roof that rises from the fascia (gutter) up to the ridge (peak).

RAKE BOARD — The rakes of some roofs are just the bare end of the sheathing. Other roofs have a wooden trim piece or rake board, which gives a more finished appearance to the rake.

RIDGE — Where the rising sides of the roof come together.

RISE — The vertical component for determining the slope or pitch of the roof.

ROOF EDGE — A trim piece that goes over the felt and is nailed up the rake.

ROOFING CEMENT — An asphalt-based adhering and sealing compound; also known as mastic.

RUN — The horizontal component for determining the slope or pitch of a roof. Compute the roof with 1 foot or 12 inches of horizontal measurement. Measure the distance up or down at the 12-inch horizontal distance to get the rise. If the run on your roof measures over 12 inches and the rise is 4 inches, you have a 4/12 roof.

SADDLE ROOF — Roofers refer to this as an "up-and-over" roof. The front roof has a straight, unbroken pitch up the the ridge and so does the rear roof. The front and rear roof segments often have the same measurements, but don't assume this.

"SHINGLE EATER" — A brand name for a shingle tear-off tool with a back-saving bent metal handle.

SHINGLE STRIPPER — Generic name for a tool designed to remove old shingle roofing. Most have notched teeth in the blade edge and a pipe welded on the heel to act as a fulcrum. Some roofers still call these "shovels" or "spades" because these are the tools that shingle strippers derived from.

SIMPLEX NAILS — An increasingly general term for nails that come with 1 inch round or square "washers." The wide washers help hold the paper in place on the sheathing.

SKIRT FLASHING — Another name for the counter flashing or finished flashing that goes over top of upper and lower flashing pieces on a chimney or skylight or over step flashing.

SLOPE — Upward or downward slant or incline.

SOFFIT — The horizontal area beneath the roof overhang over an exterior wall. The soffit is the area under the eave between the fascia and the exterior wall.

STARTER COURSE — The "upside down" course of shingles laid along the fascia.

STEP FLASHING — Pieces of uniformly bent metal which are laid under each piece of shingle tying into a wall or other obstruction.

SWING TACKER — Also called a staple hammer. A heavy-duty stapler which drives the staples when the head mechanism strikes down on the sheathing.

UP-AND-OVER — A slang description for a saddle roof where the two straight sides meet at a single ridge.

VALLEY — Where two roofs coming from different horizontal directions meet.

VENT FLASHING — The treatment of a plumbing vent. Vent flashing may be the metal-based neoprene collar or the all-lead type.

Index

Other Books of Interest

Home Construction/Repair

The Art of the Stonemason, $14.95
Building & Restoring the Hewn Log House, $18.95
The Complete Guide to Understanding and Caring for Your Home, $18.95
The Complete Guide to Home Automation, $16.95
The Complete Guide to Home Security, $14.95
The Complete Guide to Landscape Design, Renovation, and Maintenance, $14.95
The Complete Guide to Lumber Yards and Home Centers, $5.95
The Complete Guide to Barrier-Free Housing: Convenient Living for the Elderly and the Physically Handicapped, $14.95
The Complete Guide to Decorative Landscaping with Brick and Masonry, $11.95
The Complete Guide to Remodeling Your Basement, $14.95
The Complete Guide to Painting Your Home: Doing it the Way a Professional Does, Inside and Out, $5.95
The Complete Guide to Home Plumbing Repair and Replacement, $16.95
The Complete Guide to Log & Cedar Homes, $16.95
The Complete Guide to Four-Season Home Maintenance: How to Prevent Costly Problems Before They Occur, $18.95
The Complete Guide to Manufactured Housing, $14.95
The Complete Guide to Contracting Your Home: A Step-By-Step Guide for Managing Home Construction, 2nd Ed., $18.95
The Complete Guide to Residential Deck Construction, $16.95
Fireplace Designs, $14.95
Get the Most for Your Remodeling Dollar, $7.95
The Home Buyer's Inspection Guide, $12.95
Home Improvements: What Do They Cost, What Are They Worth?, $16.95
So You Want to Build a House: How to Be Your Own Contractor, $14.95

Woodworking

The Art of Fine Furniture Building, $16.95
Basic Woodturning Techniques, $14.95
Blizzard's Book of Woodworking, $22.95
Building Fine Furniture From Solid Wood, $24.95
The Complete Guide to Restoring and Maintaining Wood Furniture & Cabinets, $19.95
The Good Wood Handbook, $16.95
Make Your Woodworking Pay for Itself, $16.95
Measure Twice, Cut Once, $18.95
Pocket Guide to Wood Finishes, $16.95
The Woodworker's Source Book, $19.95

Business & Finance

Becoming Financially Sound in an Unsound World, $14.95
College Funding Made Easy: How to Save for College While Maintaining Eligibility for Financial Aid, $5.95
The Complete Guide to Buying and Selling Real Estate, $11.95
The Complete Guide to Buying Your First Home, $16.95
Homemade Money: How to Select, Start, Manage, Market & Multiply the Profits of a Business at Home, $19.95
How to Run a Family Business, $14.95
How to Sell Your Home When Homes Aren't Selling, $16.95
How to Succeed as a Real Estate Salesperson: A Comprehensive Training Guide, $14.95
Legal Aspects of Buying, Owning, and Selling a Home, $12.95
Mortgage Loans: What's Right for You?, $14.95
People, Common Sense, and the Small Business, $9.95
Rehab Your Way to Riches: Guide to High Profit/Low Risk Renovation of Residential Property, $14.95
Single Person's Guide to Buying a Home: Why to Do It and How to Do It, $14.95
The Small Business Information Source Book, $3.95
Small Businesses That Grow and Grow and Grow, 2nd Ed., $4.95
Stay Home and Mind Your Own Business, $12.95
The Student Loan Handbook: All About the Stafford Loan Program and Other Forms of Financial Aid, 2nd Ed., $7.95
Success, Common Sense and the Small Business, $11.95
Surviving the Start-Up Years in Your Own Business, $7.95
Tradesmen In Business: A Comprehensive Guide and Handbook for the Skilled Tradesman, $14.95

Sports/Coaching

The Art of Doubles: Winning Tennis strategies, $14.95
Baseball Chronicles: An Oral History of Baseball Through the Decades, $16.95
Baseball Fathers, Baseball Sons: From Orator Jim to Cal, Barry, and Ken . . . Every One a Player, $13.95
Baseball's All-Time Dream Team, $12.95
The Complete Guide & Resource to In-Line Skating, $12.95
The Downhill Skiing Handbook, $17.95
Intelligent Doubles: The Sensible Approach to Better Doubles Play, $9.95
Intelligent Tennis, $9.95
The Joy of Walking: More Than Just Exercise, $9.95
The Name of the Game: How Sports Talk Got That Way, $8.95
Never Too Old to Play Tennis . . . And Never Too Old to Start, $12.95
A Parent's Guide to Coaching Football, $7.95
A Parent's Guide to Coaching Baseball, $7.95
A Parent's Guide to Coaching Basketball, $7.95
The Parent's Guide to Coaching Hockey, $8.95
The Parent's Guide to Teaching Skiing, $8.95
A Parent's Guide to Coaching Tennis, $7.95
A Parent's Guide to Coaching Soccer, $8.95
Spinning: A Complete Guide to the World of Cycling, $14.95
The Scuba Diving Handbook: A Complete Guide to Salt and Fresh Water Diving, $19.95
Underwater Adventures: 50 of the World's Greatest!, $19.95

For a complete catalog of Betterway Books write to the address below. To order, send a check or money order for the price of the book(s). Include $3.00 postage and handling for 1 book, and $1.00 for each additional book. Allow 30 days for delivery.

Betterway Books
1507 Dana Avenue, Cincinnati, Ohio 45207
Credit card orders call TOLL-FREE
1-800-289-0963

Quantities are limited; prices subject to change without notice.